Bar code in 2 pages.

Bar code in 2 pages.

# Structural Failure in Residential Buildings

## Volume 4   Internal Walls, Ceilings and Floors

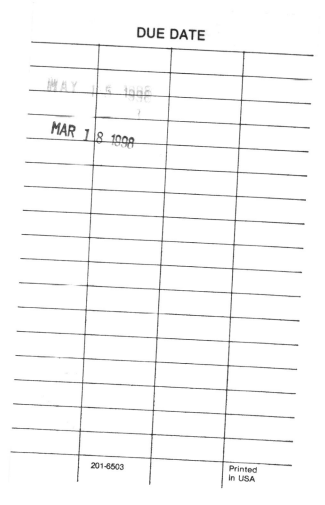

# Structural Failure in Residential Buildings

## Volume 4
### Internal Walls, Ceilings and Floors

Erich Schild
Rainer Oswald
Dietmar Rogier
Hans Schweikert

Illustrations by Volker Schnapauff

**GRANADA**
London Toronto Sydney New York

Granada Publishing Limited – Technical Books Division
Frogmore, St. Albans, Herts AL2 2NF
and
3 Upper James Street, London W1R 4BP
Suite 405, 4th Floor, 866 United Nations Plaza, New York, NY 10017, USA
117 York Street, Sydney, NSW 2000, Australia
100 Skyway Avenue, Rexdale, Ontario M9W 3A6, Canada
61 Beach Road, Auckland, New Zealand

Copyright © Granada Publishing, 1981

British Library Cataloguing in Publication Data

Structural failure in residential buildings.
  Vol. 4: Internal walls, ceilings and floors
  1. Dwellings – Defects
  I. Schild, Erich II. Schnapauff, Volker
  690'.8      TH4812        80-41704

  ISBN 0 246 11479 7

First published in 1979 in the Federal Republic of Germany
by Bauverlag GmbH, Wiesbaden and Berlin
English edition first published in Great Britain
1981 by Granada Publishing

Distributed in the United States of America by Nichols Publishing,
P.O. Box 96, New York, NY 10024 (212 580–8079)  ISBN 0–89397–100–6

Translated from the German by TST Translations

Typeset by V & M Graphics Ltd, Aylesbury, Bucks.
Printed and bound in Great Britain by
William Clowes (Beccles) Limited, Beccles and London

Granada®
Granada Publishing®

This publication is the result of the research project 'Problems of structural failure – prevention of structural damage in residential buildings', commissioned by the Ministry of the Interior (Residential Buildings Department) of North Rhine-Westphalia and carried out at the Technical University of Aachen, Faculty of Building Construction III – Building Science and Problems of Structural Failure (senior professor Prof. Dr.-Ing. Erich Schild).
Research team:

| | |
|---|---|
| Erich Schild | Prof. Dr.-Ing. |
| Rainer Oswald | Dipl.-Ing. |
| Dietmar Rogier | Dipl.-Ing. |
| Volker Schnapauff | Dipl.-Ing. |
| Hans Schweikert | Dipl.-Ing. |
| | |
| Eva Angelsdorf | cand. arch. |
| Rainer Dörter | cand. arch. |
| Elisabeth Peck | cand. arch. |
| Frank Samol | cand. arch. |

Norma Gottstein
Gunda Hoppe

## Preface to English edition

The number of failures in applied finishes to floors, walls and ceilings has increased rapidly in buildings through inadequate understanding of damp penetration, thermal conditions, lightweight structural construction and through poor site supervision. This book analyses the problems facing the designer and recommends solutions based on intensive on-site research to applied coverings. Although the book is primarily concerned with residential buildings, much of the technical information discussed – especially with regard to waterproofing floors and internal walls for utility rooms and bathrooms in dwellings – is invaluable for larger buildings such as leisure centres, swimming pools and educational buildings.

As in the three published volumes of this series *Structural Failure in Residential Buildings*, the Continental building practice of providing basements is adhered to and so the UK practice of inserting an edge thermal barrier under the ground floor slab (especially with underfloor heating) is omitted. However, the technical competency and thorough-ness of the research provide a wealth of information for the building design team and site personnel who are at present so committed to raising the standards of finishes in the completed buildings.

For UK practice it is advisable to relate the principles described in this volume to the current British Standards and Codes of Practice when applying them to site problems, as the findings themselves are pertinent but some of the minimum dimensions referred to are less than those accepted in the UK.

The English language bibliography only emphasises practical reference material; general reading matter is not included.

John G. Roberts
School of Architecture, University of Wales Institute of Science and Technology
Cardiff, 1981

# Contents

# Preface

Like the previous three volumes in this series *Structural Failure in Residential Buildings*, this volume is based on the findings of a research project carried out on behalf of the Ministry of the Interior, North Rhine-Westphalia. The project entailed a wide-ranging survey of actual cases of structural damage (the results of which were published separately*).

Whereas the first three volumes dealt with external components of residential buildings, Volume 4 is concerned with internal structures and finishes. The internal components of buildings cover a wide range of elements which differ considerably in terms of design, function and materials used. Thus, in order to prevent the subject area from becoming too complex, it was decided at the survey stage to exclude damage to mechanical installations, and so the results largely concern load-bearing and non-load-bearing internal walls, ceilings and floors, and the surface finishes of walls and ceilings. Some of the results obtained will be mentioned briefly here, since they largely determined the specific aim, emphasis and arrangement of the book.

Over half of the failures (55.9%) covered by the survey were connected with the cross section of walls and ceilings, and over one-third (37.1%) with surface finishes, including floors. Cracks resulting from deformation or elongation of the structure (and the provision for this at design stage is often inadequate) represent by far the most common source of failure in internal wall and slab cross-sections. Since it is the work of the architect and the structural engineer to ensure that the junction of components will not be liable to failure, approximately 86% of all these failures were considered as design errors. Thus, recommendations for the avoidance of this type of failure should be aimed primarily at the designers.

In examples of surface finishes to ceilings, walls and floors, the most common types and/or causes of failure were separation, blistering and cracking as a result of poor base adhesion, or the incorrect choice or application of materials. Most of these failures (66%) were attributable to construction faults – thus, recommendations for the avoidance of these defects should be aimed primarily at the site contractor.

Consequently, the aim of this book is not merely to discuss a wide range of different structural components, but also to present constructional relationships applicable to all types of component and to highlight particular construction details, thus providing recommendations for various interested groups within the building industry. Indeed, this very objective is a reflection of the complexity of designing and constructing buildings. On the one hand, it makes it possible to draw the architect's or structural engineer's attention to the practical construction problems involved and on the other hand shows the individual craftsmen involved – contractors responsible for the shell of the building, plasterers, floor screeders, tilers, floor finishers,

etc. – exactly what contribution their own individual trades make towards the finished building.

The complexity of the area under consideration makes it necessary to restrict the structures and problem areas to be discussed to the main types of failure which have been established. Thus, this book does not deal with every conceivable type of wall and floor covering; wall finishes (paints, wallpaper), soft coverings and stairs are not discussed because of the small number of failures which the survey brought to light in these examples. This book therefore does not represent a comprehensive building manual on internal components. Indeed, this was never the intention. Instead, priority is given to preventing those failures which most frequently occur.

This book has been arranged as a source of reference in the design, checking and building of structural components. In order to make it as clear as possible and to permit simple location of particular subjects, a strict layout was adopted.

The three types of structural component:

A – internal walls
B – ceilings
C – floors

are discussed in completely separate sections. Each of the main chapters is subdivided into problem areas. The descriptions of the problem areas associated with the typical cross sections of load-bearing and non-load-bearing walls and ceilings also include in each example details of the particular surface finishes applied to them. The chapter dealing with floors is subdivided into floating and bonded finishes, wearing coat and points of detail.

The treatment of the individual defects associated with the various problem areas was purposely not restricted to notes on how to remedy them. Instead, the recommendations for preventing defects are in each case preceded by a description of the damage which makes them necessary. There then follows a description of the relationship to soil mechanics, building physics and materials and building techniques that emerged from an analysis of the damage and required consideration. For only by providing an insight into the need for formal recommendations will it be possible to change methods employed in the past. Moreover, by providing justifications for the recommendations made, it will be possible to modify these recommendations to suit constantly changing situations. In this particular volume, the description of these inter-relationships is extensive so as to take account of the complex problems involved and the differing levels of knowledge of the individual interested parties. Consequently, in order to make the information more readily accessible, the main points are summarised in lists. Bibliographical notes are also provided for the reader who is interested in studying a particular problem area in greater depth.

The illustrations which accompany each defect show examples of failure and possible construction solutions. The solutions which are suggested are not intended to be the 'correct' alternative in each case; instead, the negative and positive examples should serve to illustrate the text as

* Schild, E.; Oswald, R.; Rogier, D.; Schweikert, H.: Bauschäden im Wohnungsbau, Teil VIII, Bauschäden an Innenbauteilen, Ergebnisse einer Umfrage unter Bausachverständigen, ILS Dortmund, Band 3.023, Verlag Wingen, Essen 1979.

## RESEARCH PROJECT
## PROBLEMS OF STRUCTURAL FAILURE – PREVENTION OF STRUCTURAL DAMAGE IN RESIDENTIAL BUILDINGS

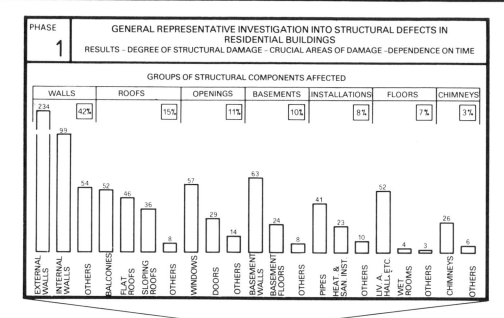

**PHASE 1**

GENERAL REPRESENTATIVE INVESTIGATION INTO STRUCTURAL DEFECTS IN RESIDENTIAL BUILDINGS
RESULTS – DEGREE OF STRUCTURAL DAMAGE – CRUCIAL AREAS OF DAMAGE –DEPENDENCE ON TIME

GROUPS OF STRUCTURAL COMPONENTS AFFECTED

| WALLS | ROOFS | OPENINGS | BASEMENTS | INSTALLATIONS | FLOORS | CHIMNEYS |
|---|---|---|---|---|---|---|
| 42% | 15% | 11% | 10% | 8% | 7% | 3% |

Bar values:
- EXTERNAL WALLS: 234
- INTERNAL WALLS: 99
- OTHERS: 54
- BALCONIES: 52
- FLAT ROOFS: 46
- SLOPING ROOFS: 36
- OTHERS: 8
- WINDOWS: 57
- DOORS: 29
- OTHERS: 14
- BASEMENT WALLS: 63
- BASEMENT FLOORS: 24
- OTHERS: 8
- PIPES: 41
- HEAT. & SAN. INST.: 23
- OTHERS: 10
- LIV. A. HALL, ETC.: 52
- WET ROOMS: 4
- OTHERS: 3
- CHIMNEYS: 26
- OTHERS: 6

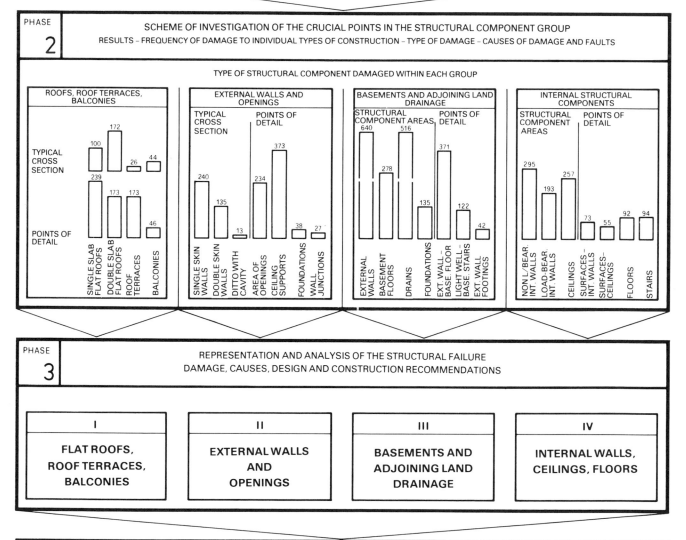

**PHASE 2**

SCHEME OF INVESTIGATION OF THE CRUCIAL POINTS IN THE STRUCTURAL COMPONENT GROUP
RESULTS – FREQUENCY OF DAMAGE TO INDIVIDUAL TYPES OF CONSTRUCTION – TYPE OF DAMAGE – CAUSES OF DAMAGE AND FAULTS

TYPE OF STRUCTURAL COMPONENT DAMAGED WITHIN EACH GROUP

### ROOFS, ROOF TERRACES, BALCONIES
TYPICAL CROSS SECTION / POINTS OF DETAIL
- SINGLE SLAB FLAT ROOFS: 100 / 239
- DOUBLE SLAB FLAT ROOFS: 172 / 173
- ROOF TERRACES: 26 / 173
- BALCONIES: 44 / 46

### EXTERNAL WALLS AND OPENINGS
TYPICAL CROSS SECTION / POINTS OF DETAIL
- SINGLE SKIN WALLS: 240
- DOUBLE SKIN WALLS: 135
- DITTO WITH CAVITY: 13
- AREA OF OPENINGS: 234
- CEILING SUPPORTS: 373
- FOUNDATIONS: 38
- WALL JUNCTIONS: 27

### BASEMENTS AND ADJOINING LAND DRAINAGE
STRUCTURAL COMPONENT AREAS / POINTS OF DETAIL
- EXTERNAL WALLS: 640
- BASEMENT FLOORS: 278
- DRAINS: 516
- FOUNDATIONS: 135
- EXT. WALL – BASE. FLOOR: 371
- LIGHT WELL – BASE. STAIRS: 122
- EXT. WALL FOOTINGS: 42

### INTERNAL STRUCTURAL COMPONENTS
STRUCTURAL COMPONENT AREAS / POINTS OF DETAIL
- NON L/BEAR. INT. WALLS: 295
- LOAD-BEAR. INT. WALLS: 193
- CEILINGS: 257
- SURFACES – INT. WALLS: 73
- SURFACES – CEILINGS: 55
- FLOORS: 92
- STAIRS: 94

**PHASE 3**

REPRESENTATION AND ANALYSIS OF THE STRUCTURAL FAILURE
DAMAGE, CAUSES, DESIGN AND CONSTRUCTION RECOMMENDATIONS

| I | II | III | IV |
|---|---|---|---|
| FLAT ROOFS, ROOF TERRACES, BALCONIES | EXTERNAL WALLS AND OPENINGS | BASEMENTS AND ADJOINING LAND DRAINAGE | INTERNAL WALLS, CEILINGS, FLOORS |

**AIM**

APPLICATION TO BUILDING PRACTICE IN ORDER TO REDUCE STRUCTURAL DAMAGE BY THE AVOIDANCE OF DEFECTS IN DESIGNING AND CONSTRUCTION

Schematic representation of stages of work

typically as possible. With this in mind, the illustrations purposely avoid giving standard details, but instead show one of several possible solutions. In the majority of the negative examples, it is not the aim to comment on the basic impracticability of the materials and products shown; similarly, with the positive examples, there is no question of suggesting that a satisfactory solution is possible only if the materials and products illustrated are used. However, once the conditions of each particular example and the recommendations derived from them are understood, it is then quite possible to check that the materials, structural components and constructions offered are suitable for the one in question.

The construction recommendations are derived from the damage which was established from an extensive evaluation of literature and product information, and from the authors' wide experience as government consultants. However, in view of the wide variety of internal components which are not covered by standards, and because of the numerous design variations, it is not possible to determine the present 'State of the Art' in the construction industry from the provisions of these standards. Despite this broad initial basis and careful preparation, the formulation of positive design and construction proposals forces one to decide which of the numerous current designs actually represent the 'State of the Art'. Indeed, in the present report, the main criterion for use as a basis for decisions was that of achieving damage-free construction over prolonged periods and even under unfavourable construction and load conditions. We did, however, feel it necessary to modify this principle when dealing with certain types of structural damage, since on the one hand their extent and effect on the inside of the building is in some instances only slight and would in any event be made good during periodical maintenance, whilst on the other hand, measures for the total elimination of damage would be disproportionately expensive. Light partition walls built on a wide-span reinforced concrete ceiling slab (see B1.1.2) are a typical example. In such examples, it would seem to be reasonable, for the sake of cost savings, to accept a design with a somewhat lower degree of reliability in terms of this type of damage. The aim of this book is to present the results of research in a form suitable for direct practical application. Not until the results of practical work have been understood and implemented can it be said that the present work has achieved its aim of preventing structural damage. The wide distribution and the great interest shown in the recommendations published in the last few years for the prevention of structural failure in flat roofs, roof terraces and balconies, external walls and openings, and in basements and adjoining land drainage seem to confirm the effectiveness of this means of presenting the information. However, an active exchange of experiences in the application of the results is desirable. We would therefore welcome critical appraisal.

Erich Schild
Rainer Oswald
Dietmar Rogier
Hans Schweikert

# Key to symbols used in the illustrations

Note: The illustrations represent principles and are not always to scale. In all illustrations, the outside of the structure is always represented by the left-hand (cross sections) or the lower side (outlines) of the picture.

Concrete, reinforced/
non-reinforced

Brickwork (also lightweight
blocks, insulation blocks, etc.)

Water-proof concrete

Water-proof render,
water-proof screed

Aerated concrete

Plaster, precast slabs

Screeds

Thermal insulation

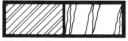
Timber/natural stone

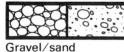
Gravel/sand

Joint compounds

Filling material, building debris

Reinforced building paper
sealing courses

Vapour barrier

Separating/plastic film

Thin building paper

Equalising layer,
spot adhesion

Continuous adhesive
layer

Compressed gravel layer

Sand layer

Rendering/reinforcement

Sealing coating (e.g. in two
layers)

Priming coat, adhesion
base

Coating

Sealing compound

Damp impregnation

Filter mat

Drainage slab (plastic)

▼GW

Standing water/ponding

Penetrating damp

⊃ ○ ○ ○ ○ ○ ○ ○ ▷
Water vapour diffusion

Damp in structural component

Emerging damp, mildew, drops
of water, soiling, etc.

Direction of ventilation

Soil, solid ground

✳ ✳
Snow

➤ ➤ ➤ ➤
Surface water

Rain

Acoustic bridging

13

# Problem: Non-load-bearing internal walls

Non-load-bearing internal walls are generally light partition walls of slight thickness (usually not more than 115 mm) which do not have any load-bearing or bracing function and which may not be exposed to any vertical loading other than their own dead weight. Moreover, since they in turn must not place any unnecessary load on other structural components, they are made as light as possible, either as bonded-brick or slab walls, or as pre-fabricated walls: framed stud partition or modular partitions. At weights of 150 kg/m², they can be regarded as a uniform addition to the live load.

However, the capacity of non-load-bearing walls to absorb, without cracking, the forces imposed on them as a result of the deformation of adjacent structural components – especially ceiling deflection – varies according to their design and the way in which they are connected (rigid, sliding or flexible) to adjoining structural components, such as walls, columns, ceilings and floors, and according to the size and distribution of openings (including ducts, conduits, and pipes etc). There may also be considerable differences in the deformation of these light partitions (creep, shrinkage, thermal buckling) compared with that of adjacent and generally load-bearing walls made of stronger and denser materials (bricks, and various blocks) as a result of the different movement characteristics of the materials used. This can be expected if the bricks are not fully cured when they are laid, or if they subsequently become wet and then dry out too quickly when the heating system becomes operational.

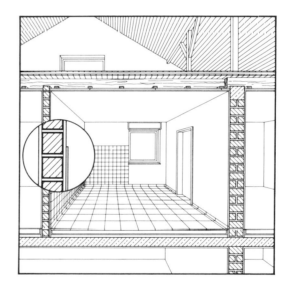

Thus, the cracking damage found in the cross sections of non-load-bearing walls was largely the result of the deformation of adjacent structural components (ceiling deflections), of the deformation of the partitions themselves and at the connecting structures which failed to take these deformations into account. With a few exceptions (sound-proofing defects in double skin walls), damage was limited exclusively to solid non-load-bearing internal walls of frame and block construction. As a result, we shall only be discussing the processes which result in the failure of this type of wall. Failure in plastered and tiled frame constructions is dealt with in section A 2.1.

Internal walls

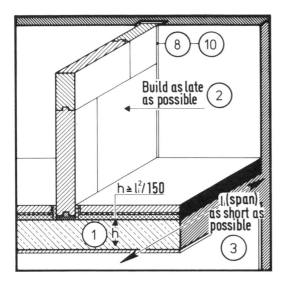

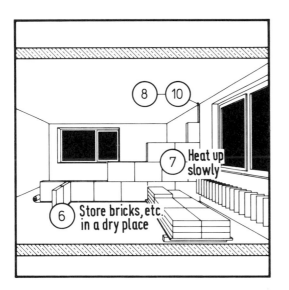

**1** The structure supporting partition walls must be as resistant to bending as possible. The recommended thickness of floor slabs must be no less than $li^2$ (span$^2$)$/150$. Specific measures should minimise creep and shrinkage deformation, which aggravates deflection (see A 1.1.2).

**2** Partition walls should be built and plastered as late as possible after the formwork to the floor slabs has been struck (see A 1.1.2).

**3** Unnecessarily large span widths between supporting structures (greater than approx. 7 m) should be avoided wherever possible if lightweight brick partition walls are to be placed on the floor slab (see A 1.1.2).

**4** If major deflection is to be expected at some later stage, the partition walls on all floors should either be self-supporting or should be of flexible construction (e.g. light framed partitions) (see A 1.1.2).

**5** Deformation behaviour should be a major criterion in the selection of the materials and constructions used for partition walls. Differential deformation in the wall structure itself and at points of connection should be as small as possible, or suitable provision should be made for expansion (see A 1.1.3).

**6** Materials should only be used if they have dried out for a sufficient period and have shrunk. Concrete materials should be stored on site in a dry place (see A 1.1.3).

**7** After the walls have been completed, they should dry out evenly by gradually warming the building and by regular ventilation (A 1.1.3).

**8** If the inclusion of high-shrinkage material is unavoidable, the damage caused by shrinkage distortion should be reduced to acceptable levels by the structural measures employed (the provision of joints to reduce continuous lengths, together with bracing if necessary; connection to adjacent structural components by means of slots, grooves, or other movement joint methods) (see A 1.1.3 and A 1.1.4).

**9** When using high-shrinkage materials (lightweight concrete, sand-lime bricks, aerated concrete) details should be obtained from the suppliers as to the extent of this shrinkage (see A 1.1.3).

**10** The connection to adjacent structures of long ($\geq$ 4 m) and high ($\geq$ 3 m) non-load-bearing internal walls of thin single skin construction, which may be subject to pressure, should be designed to accommodate slight movement. However, this connecting structure should provide the walls with sufficient resistance to lateral loads (e.g. impact, vibration, door slamming, etc.), but subsequent provision of chases and openings should be avoided (see A 1.1.4).

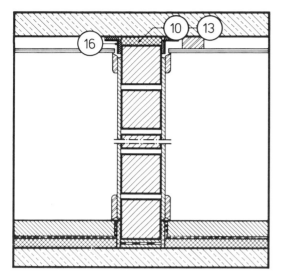

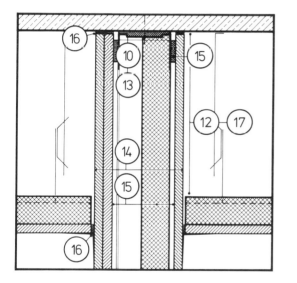

**11** The design of lightweight walls must be such that light applied loads (e.g. wall cupboards weighing $\leq 0.4$ kN/m) can be directly attached to them at any point (see A 1.1.4).

**12** Lightweight internal walls should continue through the suspended ceiling area to the supporting structure, or, provided there is sufficient stability (e.g. by connecting the walls to sufficiently rigid formwork of suitable size), adequate provision must be made for air-borne sound insulation in the air space created (see A 1.1.4 and A 1.1.5).

**13** If deflection and, particularly, elongation is to be expected in ceiling slabs, non-load-bearing internal walls should be connected to them by means of movement joints. The connecting joint must be of sufficient width that no rigid bond can be formed, even at a later stage (see A 1.1.5).

**14** If increased requirements are placed on non-load-bearing or lightweight partition walls in terms of sound insulation $\geq 48$ dB, or air-borne sound insulation level $\geq -3$ dB, or if heavy loads cannot be placed on the upper floor slab, or if relatively major distortion is anticipated in the structural components which support the walls (ceiling deflections), the walls should consist of two or more skins and should be made of framed construction (e.g. timber stud or metal frame) (see A 1.1.5).

**15** In frame constructions, there should be no rigid connection between the two framed skins. They should, therefore, wherever possible be fitted to independent frames made of timber or to pressed metal channel sections. However, in examples of a single stud frame, the two external skins should at least be fixed with soft spacing pieces installed as far apart as possible ($\geq 800$ mm). Depending on the type of skin finish and the materials used (plasterboard, chipboard, lightweight woodwool slabs) it may be necessary to use supporting battens to reduce deflection and warping.

The space between the skins should be as wide as possible ($\geq 100$ mm), and if the internal surfaces of the skins are not porous (e.g. lightweight woodwool slabs) the cavity should be filled with sound-absorbing materials between the studding. In addition, varying panel thicknesses in the two skins, or two-coat finishings, have a favourable effect on the soundproofing characteristics of the wall (see A 1.1.5).

**16** All joints and edge connections must be carefully sealed for sound insulation purposes (see A 1.1.5).

**17** Areas above suspended ceilings must either be partitioned off with bulkheads or must be sufficiently soundproofed by inserting soft damping materials over a width of at least 1 m around the point of connection (see A 1.1.5).

Internal walls
Non-load-bearing internal walls

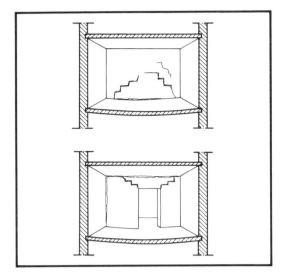

Non-load-bearing internal walls were particularly prone to cracking where they were erected on ceiling/floor slabs or beams which subsequently suffered severe deflection.

Typically, cracking took the following forms: diagonal cracks in the corners, cracks at junctions with walls and columns, ceilings or floors, and in isolated examples, virtually parallel vertical cracks mostly in the lower parts of the walls, or horizontal cracks about half way up. This type of cracking affected only jointed and bonded brick and block walls.

**Points for consideration**

- The deformation (e.g. deflection) of structures which support walls produces bending pressure, tensile and shearing stresses in the walls which rest on them. Moreover, if the acceptable stresses are exceeded, cracking results, since mineral building materials, especially brick-work, are able to absorb only very slight stresses, particularly bending and shearing stresses.

- Damage-inducing deformation in the structures which support walls may, among other things, be caused by ceiling/floorslabs which are too thin to resist bending, irregular and uneven settling of the building on site, or by the use of brickwork with different deformation characteristics for the external and internal walls.

- The amount of deformation caused by the dead load of the ceiling slab and wall and the imposed load which occurs before or immediately after the walls have been erected does not have any effect on the risk of cracking since, until the jointing mortar has fully hardened, the brickwork is initially able to absorb deformation within certain limits, and since any cracks which appear before plastering can be covered. Alternatively, those types of deformation (creep and shrinkage) which can occur up to five years after completion of the structure, depending on the composition of the concrete mix and the way in which it is worked, compressed and finished, do have a significant bearing on the susceptibility of a structure to cracking.

- Experience has shown that no cracking can be expected in the cross-section of a wall where the width between supports is small ($\leq$ 4500 mm) and where deflection is slight ($\leq$ 15 mm), since within these stress limits, a brickwork wall is self-supporting, or it forms a 'supporting arch'. Nevertheless, cracking can occur in the middle of the wall where it connects with the floor. However, the risk of cracking rises as span widths and deflection increase, since, as a result of the wall supports spreading, the dead load to be supported by the wall panel finally becomes greater than the tensile strength of the brickwork and so wall sections are able to settle on the ceiling slab. Under these circumstances, we can expect horizontal cracks in the lower parts of the wall and diagonal cracks pointing towards the corners of the support.

- If the brickwork wall is exposed to deformation as a result of evenly deflected ceiling slabs built on top of one another in a multi-storey building, or as a result of sagging ceilings there is a risk of cracks forming across the shortened diagonal lines of the wall, or – particularly in examples of wide span widths and where the wall panels are exposed to additional loads – there may be several vertical cracks side by side, similar to those found in a reinforced concrete balcony which sags.

- Deflected ceiling/floor slabs place a load on the walls erected beneath them and may result in mortar being forced

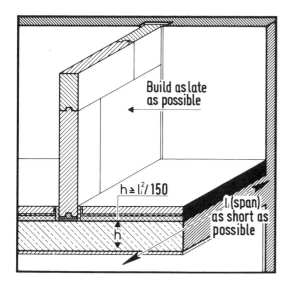

out from the joint between the ceiling and the wall and may cause increased sound transmission by conducting sound through the length of the loaded partition walls which are connected directly to the ceiling.

– Particularly where door openings interrupt the wall area, the panel and supporting arch effects of the wall are destroyed and cracks appear along the ceiling in the direction of the corners of the door opening and at lateral wall junctions so that the individual sections of the wall can follow the deflection.

– The absence of cracking depends not only on the extent of the deflection as a function of span width but also on the overall extent of the cracking. The extent of deflection can therefore be reduced only within limits by reducing the 'bending slenderness' of ceiling slabs.

– Moreover, deflection is, above all, influenced by the inherent deformation of the concrete (creep and shrinkage), which is largely a result of factors associated with its composition (grading, water/cement ratio) and of the way in which it is treated after laying (keeping it damp, erection of temporary supports).

– In addition to comply with the recommendation that the required thickness $= li^2$ (span$^2$)/150, ceilings which support partition walls would have to be uneconomically thick.

– Horizontal reinforcement in walls can only reduce vertical cracks resulting from bending. It will not prevent diagonal shear cracks.

### Recommendations for the avoidance of defects

● The structure supporting partition walls must be as resistant to bending as possible. The recommended thickness of floor slabs must be no less than $li^2$ (span$^2$)/150. Specific measures (grading, water/cement ratio, keeping damp after erection, erection of temporary supports, etc) should minimise creep and shrinkage deformation which aggravates deflection.

● Partition walls should be built and plastered as late as possible after the formwork to the floor slabs has been struck.

● Unnecessarily large span widths between supporting structures greater than 7 m should be avoided wherever possible if lightweight brick partition walls are to be placed on the floor slab.

● If major deflection is to be expected at some later stage, the partition walls on all floors should either be self-supporting or should be of flexible construction (e.g. light framed partitions).

19

Internal walls
Non-load-bearing internal walls

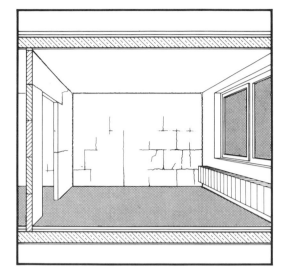

Light partition walls made of lightweight concrete, aerated concrete, or sand-lime bricks had parted from adjacent structures (slabs, load-bearing walls); in addition to these settling cracks, there were frequently cracks in the central area of the wall – mostly vertical and running parallel to one another – which partly followed the course of the joint and continued through the whole of the wall's cross-section. In examples of aerated concrete slabs bonded together on a thin bed of mortar, hairline cracks appeared immediately the heating was put into operation (and also during the internal finishing work). This cracking was more serious near radiators.

## Points for consideration

– The volume of most building materials is liable to decrease as a result of drying out. Moreover, these changes are increased the greater the content of fine particles and consequent larger initial moisture content of the structural component, and the smaller the thickness of the structural component, or the greater the surface area of the structural component in relation to its volume. This is generally the case in partition walls.

– Since internal walls themselves have a very low moisture content because of the dry atmosphere in the room, correspondingly high levels of shrinkage expansion can be expected – especially when using materials which are bonded together with water based mortars.

– The progress and extent of shrinkage expansion depend on drying conditions, i.e. as air humidity falls, temperature rises (heating) and air movements increase, while shrinking speed and the extent of final shrinkage expansion increase.

– If different materials with differing deformation characteristics are used in a building (e.g. infilling reinforced concrete frames with lightweight concrete blocks), cracking will occur at connecting joints if the shrinking stresses present are greater than the bonding forces. If the walls are rigidly connected to adjacent structural components by frictional forces, several finely distributed 'shrinkage cracks' will appear in the weakest points of the wall (joint areas and in some cases around the corners of doors) because they cannot deform when the tensile stresses become more than they can absorb.

– Materials that are not cured when used tend to result in increased shrinkage cracking, since initial shrinkage, which lasts for different times according to the material used (it takes about 3 months for up to 95% of the shrinkage to occur in lightweight concrete blocks) is still not complete. Similarly, aerated and foamed concrete are high shrinkage materials, with overall shrinkage of 2 mm per metre length. They should not be installed until shrinkage has dropped to no more than 0.5 mm per metre length.

– If materials become wet after construction, the whole shrinkage process is delayed; with some materials (e.g. stock bricks) this may result in irreversible expansion (swelling).

– Because of the relatively small load of light non-load-bearing walls (dead load) only slight creep expansion occurs, and this can be disregarded.

**Recommendations for the avoidance of defects**

● Deformation behaviour should be a major criterion in the selection of the materials and construction used for partition walls. Differential deformation in the wall structure itself and at points of connection should be as small as possible, or suitable provision should be made for expansion.

● Material should only be used if it has dried out for a sufficient period and has cured. Brick and blocks should be stored on site in a dry place.

● After the walls have been completed, they should be induced to dry out continuously by gradually warming the building and by regular ventilation.

● If the inclusion of high-shrinkage material is unavoidable, the damage caused by shrinkage distortion should at least be reduced to low levels by the structural measures employed (the inclusion of movement joints, connection to adjacent structural components by means of slots, grooves).

● When using high-shrinkage materials (lightweight concrete, sand-lime bricks, aerated concrete) details should be obtained from the suppliers on the extent of this shrinkage.

Internal walls
Non-load-bearing internal walls

If non-load-bearing internal walls were rigidly connected to adjacent structural components (wall, column, ceiling) and if compression forces were applied to the wall, or if the walls were made of high-shrinkage materials, there was frequent cracking around junctions, which at the same time had the effect of reducing the soundproofing properties of the wall. This type of damage was found only in bonded brickwork walls.

In addition to cracking, sound transmission was particularly noticeable at connections to suspend ceilings. Cracking also occurred where flexible connections had been incorrectly made.

### Points for consideration

- Rigid connecting structures with anchor bolts, dowels and steel angles etc. do not allow any movement; they thus produce stresses in the wall which lead to cracks once the tensile and shearing stresses which the wall can withstand have been exceeded.

- Sliding or flexible connections permit deformation of the wall within certain limits. The purpose of the connection is not to transmit loads from adjacent structural components to the lightweight walls, but merely to support these walls against lateral horizontal loads. However, considerable stresses can be placed on lightweight partition walls as a result of impact and applied loads (wall-mounted cupboards), as well as vibration (door slamming, chasing work). Similarly, as walls become thinner, especially in the case of chases, recesses and openings, they may have insufficient stability. The location and construction of wall openings, chases and recesses and especially those for which there are no details of permissible wall heights and lengths, are set out in the individual building standards relating to materials.

- Inadequate soundproofing as a result of the indirect transmission of sound through the air is to be expected where non-load-bearing walls are not continued and secured to the supporting structure but are connected to suspended ceilings without the provision of special sound-proofing measures (see A 1.1.5). Similarly, the damping of air-borne sound can be considerably reduced by open channels and gaps in connecting joints as a result of the above-mentioned processes.

- Storey wide slabs are prone to deformation as a result of deflection. Moreover, in the case of roof slabs (especially solid concrete flat roofs), horizontal length fluctuations may occur as a result of temperature stresses, and where there is a rigid connection between the slab and the wall, these deformations will be transmitted to the wall and will produce cracking around the junction because of the wall's low shear strength.

Internal walls
Non-load-bearing internal walls

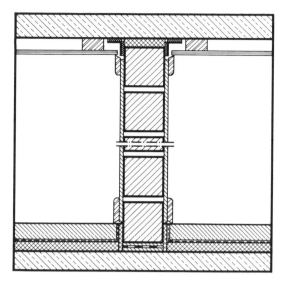

### Recommendations for the avoidance of defects

● The connection to adjacent structures of long ($\geq$ 4 m), and high ($\geq$ 3 m), non-load-bearing internal walls of particularly thin construction, which may be subject to pressure loads, should be of the sliding or flexible type (e.g. with inlaid mineral fibre or bituminous strips). However, this connecting structure should provide the walls with sufficient resistance to lateral loads (e.g. impact, vibration, door slamming, etc.). Chasing work (e.g. the subsequent provision of chases and openings) should be avoided.

● The continuous lengths of walls made of high-shrinkage material should be as small as possible, or provision should be made for a suitable number of expansion joints (if necessary in conjunction with bracing measures).

● The size of lightweight walls must be such that light applied loads (e.g. wall cupboards weighing $\leq$ 0.4 kN/m) can be directly attached to them at any point.

● Lightweight internal walls should continue through the suspended ceiling area to the supporting structure, or, provided there is sufficient stability (e.g. by connecting the walls to sufficiently rigid formwork of suitable size), adequate provision should be made for air-borne sound insulation in the air space created.

● If deflection, and particularly elongation, is to be expected in ceiling slabs, non-load-bearing internal walls should be connected to them by means of movement joints. The connecting joint must be sufficiently wide to ensure that no rigid bond can be formed, even at a later stage.

Internal walls
Non-load-bearing internal walls

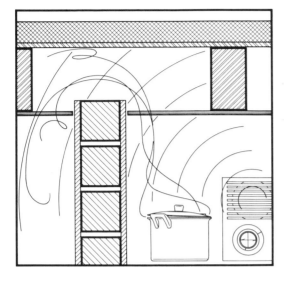

In addition to cracking in single skin, non-load-bearing internal walls, inadequate design and construction of the wall's cross section and connections, and even of multi-skin walls, were the main cause of defective damping of air-borne sound. Thus, for example, an 115 mm thick partition constructed of sand-lime bricks – rigidly connected to the side walls by a relatively stiff 20 mm thick hard foam insulation board bonded to a 95 mm pumice backing – was not able to offer the necessary level of insulation against air-borne sound of ± 0 dB for apartment partition walls. Inside apartments, considerable sound problems were experienced if the partition wall did not continue as far as the ceiling slab structure, but ended merely above the suspended ceiling (e.g. plaster board on the underside of timber joist ceiling constructions).

Double skin partition walls constructed on a stud frame construction had poor sound insulation properties as a result of the rigid connection between the two skins through their common frames.

## Points for consideration

– Lightweight single skin partition walls (e.g. a sand-lime brick wall) plastered on both sides, achieve an estimated maximum sound proofing level of approx. 48 dB (air-borne/sound insulation level = – 4 dB). Identical and better sound proofing levels can be obtained with double skin walls, whilst making savings in terms of weight.

– The soundproofing characteristics of double skin lightweight wall construction depends on the surface pliability of the individual skins which limit frequencies outside the acoustically significant frequency range, on the inherent frequency level of the wall system itself – in particular it is the width of the cavity, and the material with which it is filled, that is important – on the restraint at adjacent structural components, on the type of supporting structure, and on the way in which the surface panels are fitted to the stud frames.

– Double skin wall structures consisting of two rigid skins of approximately equal thickness and weight per unit of area are unable to improve on the soundproofing characteristics of single skin walls of the same weight if most of the sound is transmitted through a rigid, fixed-edge connection. In this example, the cavity between the skins and the cavity filling have little effect on soundproofing characteristics.

Similarly, wall structures in which the skins consist of soft, pliable panels of the same material, and of the same thickness, may have considerably poorer air-borne sound insulation characteristics than those of a solid single skin wall, because of the superimposition of resonant frequencies.

– Any suspect areas in the wall – such as cracks, non-sealed edge connections, openings, open joints, and, particularly in the case of panels which are built-in, dry (plaster board) – will reduce the level of damping of air-borne sound.

– If, in double skin lightweight partition walls, soft pliable skins are rigidly connected to common stud frames, the resultant estimated soundproofing levels are only approximately 37 dB (air-borne sound/insulation = – 15 dB) to 42 dB (air-borne sound insulation = –10 dB), depending on the type and thickness of cladding, and on the distance between the skins and the supporting frame as a result of acoustic bridging. The soundproofing effect can be improved somewhat by fitting two soft pliable panels on either side, or, if the panels are not attached to the supporting frame, over their whole surface (in strips), but they should be secured at individual points by some pliable means. In this way, estimated sound insulation levels of up to approximately 52 dB (airborne sound insulation 0 dB) can be achieved.

Internal walls
Non-load-bearing internal walls

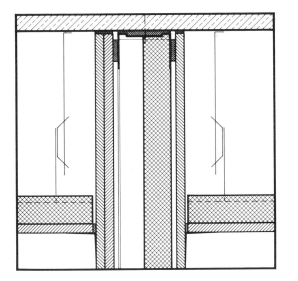

### Recommendations for the avoidance of defects

● If increased requirements are placed on non-load-bearing or lightweight partition walls (in terms of sound insulation $\geq$ 48 dB, or airborne sound insulation level $\geq$ 4 dB), or if heavy loads cannot be placed on the upper floor slab, or if relatively major distortion is anticipated in the structural components which support the walls (ceiling deflections), the walls should consist of two or more skins and should be made of framed construction (e.g. timber stud or metal frame).

● In frame constructions, there should be no rigid connection between the two framed skins. They should, therefore, wherever possible be fitted to independent frames made of timber, or to pressed metal channel-sections, or should at least be fitted with soft spacing pieces installed as far apart as possible ($\geq$ 800 mm). Depending on the type of skin finish and the materials used (plasterboard, chipboard, lightweight woodwool slabs) it may be necessary to use supporting battens to reduce deflection and warping.

The space between the skins should be as wide as possible ($\geq$ 100 mm), and if the internal surfaces of the skins are not porous (e.g. lightweight woodwool slabs) the cavity should be filled with sound-absorbing materials between the studding. In addition, varying panel thicknesses in the two skins, or two coat finishings, have a favourable effect on the soundproofing characteristics of the wall.

● All joints and edge connections must be carefully sealed for sound insulation purposes.

● Areas above suspended ceilings must either be partitioned off with bulkheads or must be sufficiently soundproofed by inserting soft damping materials over a width of at least 1 m around the point of connection.

## Problem: Load-bearing internal walls

Load-bearing internal walls are subjected to the widest variety of stresses resulting from loadings, inherent deformation and site settlement. On the one hand, with the common and widely used rigid systems or supporting structures found in today's residential buildings, the internal walls which form the support generally bear a greater proportion of the load than the external walls. Alternatively, different materials are, for constructional, rigid and economic reasons, frequently used for external and internal structural components, and there may be considerable differences in their deformation characteristics. In addition to this, there are different climatic influences (outside: sunlight, daily and annual temperature fluctuations, water vapour and precipitation; inside: largely constant 'heating systems'). In unfavourable circumstances, where there is a combination of these influences, there can be considerable differences in the expansion and settling of external and internal structural components.

In general, the external and internal walls are solidly connected to floor slabs and transverse walls, which have a bracing action; consequently, they are unable to sustain deformation freely and independently. This can lead to stress in the structural components. If the loaded internal walls are made of brickwork, there is a particularly high risk of cracking as a result of the low shearing and tensile strengths of the walls. To ensure that structures which are loaded in this way are free of cracks and of the risk of cracking, therefore requires careful investigation and determination of the possible expansion differentials within the supporting structure in addition to a precise knowledge of the materials to be used.

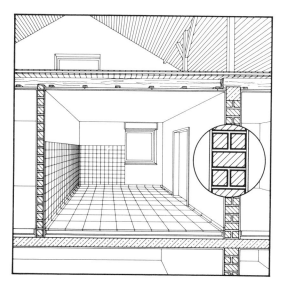

As the survey showed, cracking which continued through the whole of the wall's cross section occurred predominantly at points of connection or in the area of adjacent structural components (ceiling slab, external wall). This common type of defect and its causes will be discussed below.

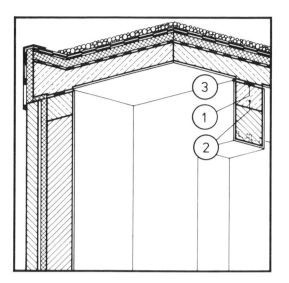

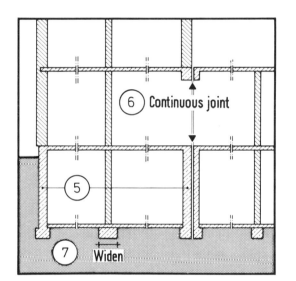

**1** Where the solid roof slab of multi-story buildings has long effective displacement lengths ($\geq$ 6 m), and if there is evidence of deformation which might create damage, it is necessary to provide a sliding support, the supports for all load-bearing walls must take the form of a sliding bearing across the whole thickness of the wall. Non-load-bearing internal walls must have flexible connections (see A 1.2.2).

**2** A reinforced concrete bracing ring beam can only be dispensed with where supports are able to offer sufficient resistance to temporary loads (e.g. as a result of wind). However, within a bed of mortar reinforced steel rods about 30 mm thick should be arranged beneath the top layer of bricks to act as a basis for the support (see A 1.2.2).

**3** Layers of plaster or wallpaper should be separated in the area of the sliding joint (see A 1.2.2).

**4** The shrinkage characteristics of the materials used for the roof slab and the brickwork should be as similar as possible (see A 1.2.2).

**5** If soil surveys reveal the possibility of major differences in the settlement of the building, the latter should be constructed as rigidly as possible (e.g. in short buildings in the form of reinforced concrete compartments), or should be capable of withstanding deformation (see A 1.2.3).

**6** Large buildings ($\geq$ approx. 30 m) should be subdivided by means of joints. Similarly, parts of the building which are of different heights should also be separated from one another by means of settlement or movement joints. The joints must continue through the whole of the building, including the foundations, and should not be filled with mortar. Settlement joints should also be provided where soil types with different load-bearing characteristics meet (see A 1.2.3).

**7** Depending on the soil type, reinforcement or compaction of the foundation soil may be suitable means of minimising differential settlement in addition to the specific choice of foundation type and size. The construction schedule (relatively slow building rate, erection of the heavy structures before the lighter ones, construction of wall infill as late as possible in the case of framed structures) can also have a favourable effect on the prevention of cracking as a result of settlement (see A 1.2.3).

**8** Any modifications to the soil which may subsequently affect load-bearing characteristics (groundwater sinking, possibly also as a result of the provision, at a later stage, of land drainage, or of the cultivation of surrounding land, and particularly as a result of the excavation of trial holes adjacent to existing buildings, or possibly with underpinning work) should only be made after careful investigation of the foundation soil and of the load-bearing capacity and strength of the structural components (foundations, slabs, walls). In some examples, it may be necessary to take suitable protection and strengthening measures (soil reinforcement, underpinning, shoring) in adjacent buildings (see A 1.2.3).

Internal walls

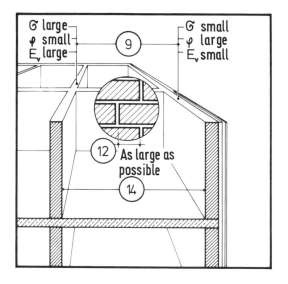

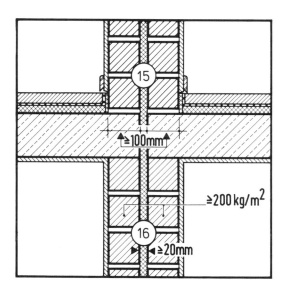

**9** In external and internal walls made of brickwork, the differential deformation caused by load and moisture should be as slight as possible. Moreover, in order to reduce differential expansion, bricks of a higher crushing strength than the structural calculations demand can be used for the internal walls and bricks of a lower crushing strength for the external walls (see A 1.2.4).

**10** Temperature differences between the internal and external wall should be reduced as far as possible, e.g. by means of double skin external walls or additional thermal insulation to the outside of the external wall (see A 1.2.4).

**11** In load-bearing partition walls, the height-to-length ratio should wherever possible not exceed approx. 1:4 in 5-storey buildings, and approx. 1:3 in 3-storey buildings (see A 1.2.4).

**12** The type of bond and the size of bricks used in the brickwork should be selected to ensure sufficient overlapping of the bond (see A 1.2.4).

**13** The bricks should be as dry as possible when supplied and laid and should be protected against becoming soaked once they have been laid (see A 1.2.4).

**14** In buildings with more than 4 storeys in particular, external and internal walls should be constructed of the same materials unless an approximate estimate of the risk of cracking has shown that the differential expansion will not have any harmful effect (see A 1.2.4).

**15** Single skin brick walls provide sufficient sound insulation as partition walls between apartments (and within a single apartment) provided that they are of sufficient density (= 480 kg/m²) and are adequately sealed (plastered on both sides) and provided that they have sufficient resistance to bending. For partition walls between houses (semi-detached or terraced houses) and for walls which separate apartments from rooms where there is a concentration of sound only double skin walls can provide the necessary sound insulation (see A 1.2.5).

**16** Double skin house partition walls must be separated by a continuous cavity across their width. This joint should if possible be 30 mm wide ($\geq$ 20 mm) and should be filled with sound insulation materials (e.g. mineral fibre insulation panels). Sound bridges caused by mortar droppings should be avoided in all instances (see A 1.2.5).

**17** If, after they have been constructed, it is found necessary to improve the sound insulation characteristics of solid dense walls additional skins which are as pliable as possible (e.g. plasterboard or lightweight woodwool panels) should be fixed by light framing to the front of the walls providing a largest possible gap between the panel and the wall. This cavity should then be filled with soft insulation material (e.g. mineral wool). Indirect sound transmission should as far as possible be avoided by separating the framing members and the panels with sound absorbing pads (ceiling, side walls) (see A 1.2.5).

**18** There should be a 'soundproof' connection between walls and the ceiling. In examples of suspended ceilings, the cavities should where possible be partitioned off or should be filled with suitable insulation materials (see A 1.2.5).

Internal walls
Load-bearing internal walls

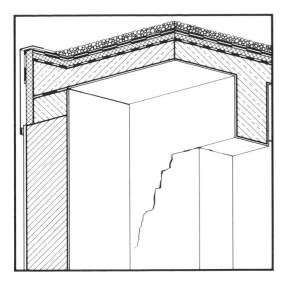

Load-bearing internal walls made of brickwork beneath solid roof slabs were particularly prone to damage where they supported these slabs. Diagonal cracks were found in partition walls and lateral walls both in multi- and single-storey buildings as a result of this. Whilst in the area of direct support, fine horizontal cracks were found in internal walls immediately beneath the slab (and sometimes this continued to two layers of bricks below), near the external walls there were open diagonal cracks whose corners pointed towards the outside. Large areas of cracking also occurred where ring beams and sliding bearings were provided only for the external walls, with the internal walls being solidly connected to the roof slabs. Similarly, at the cornices of plastered roofs, there was also cracking and especially where the ring beam was joined to the ceiling or the wall.

The flat roofs investigated consisted exclusively of reinforced concrete slabs, largely in the form of conventional 'warm roofs' with external thermal insulation, but there were some examples of 'water-proof concrete roofs' with thermal insulation on the inside. Damage in the supporting area of inner floor slabs was less common. This type of damage affected load-bearing internal walls, which formed the support for the ends of wide-span reinforced concrete slabs (e.g. staircase walls), where it formed cracks at the edges under compression.

**Points for consideration**

– Considerable differential expansion may occur between solid roof slabs and the walls erected beneath them as a result largely of shrinkage and temperature movement. The extent of this differential expansion depends among other things on the specific deformation characteristics – largely shrinkage and thermal expansion rates – of the materials used for the roof and the wall. The shrinkage characteristics of the roof can be largely influenced during construction, whilst linear thermal expansion depends on the thickness and position of the thermal insulation layer, on the position of the fixing point and on the way in which expansion joints are arranged.

– It is significant movements in the roof slab parallel to the wall surfaces which provoke tensile and shearing stresses, with excessive loads in the wall forming the roof support, in the event of contractions and extensions of the roof slab relative to the wall. These stresses can only be absorbed to a very limited extent without cracking, because there is no brickwork load imposed on the roof storey. Reference should be made to existing literature (see A 1.3.0) to determine the permissible differential expansion between the slab and the wall.

– Severe linear deformation can be expected in large, continuous, joint-free roof surfaces which may have unfavourable support (away from the centre), and in flat roofs (waterproof concrete roofs) with thermal insulation to their undersides giving no external protection against temperature fluctuations. This deformation increases proportionally to the length of the structure and the temperature difference. Similarly, the risk of cracking is at its greatest in long, continuous and retained wall panels. Shorter wall sections, with wherever possible, openings of the same height as the wall, permit relatively reduced areas more able to absorb flexural deformation.

– Walls in single-storey buildings are at greater risk than those in multi-storey buildings, since greater temperature differences occur between the basement ceiling slab (or the floor slab in buildings with no basement) and the roof slabs.

29

Internal walls
Load-bearing internal walls

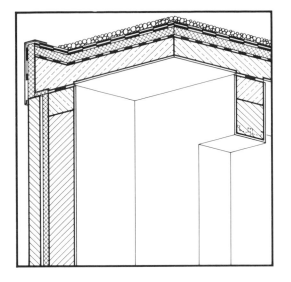

– This differential expansion frequently results in cracking a few brick courses lower than the bearing joint immediately beneath the slab, since this joint is generally more resistant to shearing stresses than the other mortar joints.

– Even despite the inclusion of reinforced concrete ring beams, damage can occur in the structural support area – especially if the sliding bearings are incorrectly constructed – if deformation can be transmitted to the brickwork as a result of an unintended shear-resistant connection between the ceiling and the walls.

– A roof slab construction in which the roof is able to slide or be displaced only on the external walls is of no use in ensuring that the internal walls will be free of cracking (nor in preventing displacement of the corners of the external walls) since the roof slab movements are also active internally.

– Compared with a fixed support, a combination of sliding bearings and ring beams represent expensive construction whose effectiveness depends largely on the material of which the bearing is made and on the care taken during construction. Therefore, a check should be carried out prior to construction in each individual example to establish whether a displacement bearing and a ring beam (e.g. when using viscosity bearings) is necessary.

– Wide-span reinforced concrete roof slabs instigate severe torsion in the end bearings, which can lead to lifting of the undersides of the slabs and to cracking and plaster spalling in the area between the bearing and load-bearing internal walls. Internal walls also frequently form such bearings (e.g. staircase walls) and cracking damage is particularly noticeable in this area, since internal bearings are generally not covered, as is the case with external wall bearings.

– Further details, particularly relating to the function and construction of displaceable and non-displaceable bearings can be found in *Structural Failure in Residential Buildings, Volume 1 – Flat Roofs, Roof Terraces, Balconies* and in *Volume 2 – External Walls and Openings*.

**Recommendations for the avoidance of defects**

● Where the solid roof slab of multi-storey buildings has long effective displacement lengths ($\geq$ 6 m), and if there is evidence of deformation which might create damage, it is necessary to provide a sliding support, the supports for all load-bearing walls must take the form of a sliding bearing across the whole thickness of the wall. Non-load-bearing internal walls must have flexible connections (see A 1.1.4).

● A reinforced concrete bracing ring beam can only be dispensed with where supports are able to offer sufficient resistance to temporary loads (e.g. as a result of wind). However, within a bed of mortar, reinforced steel rods about 30 mm thick should be arranged beneath the top layer of bricks to act as a basis for the support.

● Layers of plaster or wallpaper should be separated in the area of the sliding joint (see B 2.1.8).

● The shrinkage characteristics of the materials used for the roof slab and the brickwork should be as similar as possible.

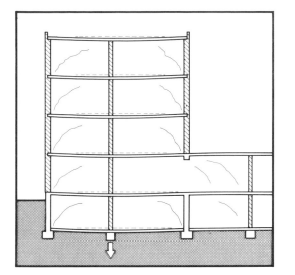

If there was relatively major settlement of the foundations of the structure of varying degrees, which would to some extent add to the differential expansion caused by the varying deformation characteristics of the different materials used (see A 1.2.4) cracking occurred not only in the external walls but also – and frequently a relatively long time after completion of the building – in the internal walls. Basically, these cracks were found to rise towards that part of the structure which had settled most, and adopted similar patterns and the same severity in all stories above. Vertical cracks appeared along connecting joints in parts of the building of different heights and with different loadings, which were not, or were inadequately, separated by a movement joint.

**Points for consideration**

– Because of the imposed load of the building, the soil beneath the foundations which supports this load becomes deformed, and if this settlement is irregular and of varying forms, there is a risk of cracking in the load-bearing brickwork. This is the case when the soil conditions differ (e.g. isolated strata of soft soil layers – such as clay, peat or loam – in the ground directly supporting the building), in the ground immediately beneath the imposed load of the building, or if the foundation soil is saturated or frozen, or if the groundwater level has been lowered.

– If the foundations for heavily loaded internal walls are too small in comparison to the foundations for the external walls, there are particularly large differences in the degree of settlement as a result of the different degrees of soil compression beneath the internal and external walls.

– In buildings composed of several sections of varying heights, settlement depressions of different depths form because of the uneven loadings.

– If sections of the building are constructed at the same time, cracks may appear (despite the provision of settlement joints) because of the increased loading on the lower parts of the building as they sink in the poor ground.

– Groundwater variations (rise or fall) also alter the soil structure, and thus the load-bearing capacity of the foundation soil. Subsequent irregular settlement which can pose a threat to the building occurs particularly when there are irregular drops in the groundwater level, for example when excavating groundwater-filled trial pits alongside existing buildings ('infill sites'). Moreover, with infill development there is a virtually unavoidable risk of cracking in the adjacent building, since the settlement depression which forms beneath the new building will extend beneath the adjacent building and result in an overhang which will cause bending and shearing stresses in the ends of the structure.

– During the course of the building work, brickwork is able to adapt to limited settlement in the foundation soil because of its initially good creep characteristics, largely without any risk of cracking. As the hardening of the jointing mortar progresses, the capacity of the brickwork to absorb deformation becomes very small, and so the risk of cracking increases, particularly in the case of settlement which takes place over a long period of time, for example with cohesive soil types.

Internal walls
Load-bearing internal walls

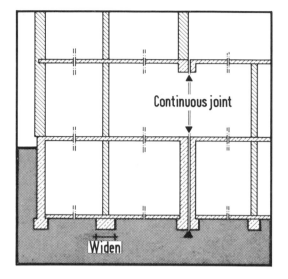

– It is not the total settlement dimensions which are of significance in the occurrence of damage, but the differential settlement. As settlement begins to form, and particularly beneath long buildings in homogeneous, soft soil types, this differential settlement can become so great that the brickwork can no longer absorb it without cracking.

– A *criterium* for evaluating the risk of cracking is the angular twisting resulting from the differences in the settlement of adjacent foundations. On the basis of various measurements, observations and investigations on this subject, it can be assumed that in reinforced concrete, framed or brickwork structures, where differential settlement is more than 1/500 (e.g. in a framed structure with brickwork infill) to 1/300 (1 = distance between two adjacent foundations) cracking must be expected. This is particularly true in buildings of low rigidity (tall, slim buildings) and for brickwork wall panels containing many door openings. In these examples it is impossible for sufficient compression to form in the settlement area, which would, by transmitting the loads to the outside, help to equalise the differences in settlement between the external and internal walls.

## Recommendations for the avoidance of defects

● If soil surveys reveal the possibility of major differences in the settlement of the building, the latter should be constructed as rigidly as possible (e.g. in short buildings in the form of reinforced concrete compartments, in loadbearing brickwork at least the basement should be made of reinforced concrete) or should be capable of withstanding deformation.

● Large buildings (> approx. 30 m) should be subdivided by means of joints. Similarly, parts of the building which are of different heights should also be separated from one another by means of settlement or movement joints. The joints must continue through the whole of the building, including the foundations, and should not be filled with mortar. Settlement joints should also be provided where soil types with different load-bearing characteristics meet (soil survey).

● Depending on the soil type, reinforcement or compaction of the foundation soil may be a suitable means of minimising differential settlement, in addition to the specific choice of foundation type and size. The construction schedule (relatively slow building rate, erection of the heavy structures before the lighter ones, construction of wall infill as late as possible in the case of framed structures) can also have a favourable effect on the prevention of cracking as a result of settlement.

● Any modifications to the soil which may subsequently affect load-bearing characteristics (groundwater sinking, possibly also the provision, at a later stage, of land drainage, or the cultivation of surrounding land, and particularly as a result of the excavation of trial holes adjacent to existing buildings, or possibly with underpinning work) should only be made after careful investigation of the foundation soil and of the load-bearing capacity and strength of the structural components (foundations, slabs, walls). In some examples, it may be necessary to take suitable protection and strengthening measures (soil reinforcement, underpinning, shoring) in adjacent buildings.

Internal walls
Load-bearing internal walls

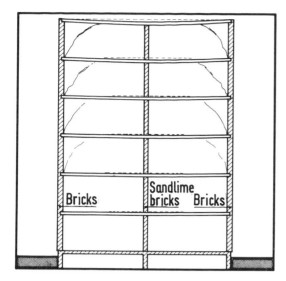

Bricks Sandlime bricks Bricks

In multi-storey buildings, partition walls made of brickwork were found to have diagonal cracks running down to the abutment adjacent to the external wall, particularly when there was a considerable difference between the deformation characteristics of the materials used for the external and internal walls (e.g. clay bricks for the outside, sand-lime bricks for the inside). Similarly when the bricks were not allowed to dry out sufficiently before laying, or if, because of load distribution, there was major differential expansion, for example between the external wall and the central wall. It was also quite noticeable that cracking, which started in the 4th or 5th storeys as fine hairline cracks, became progressively more serious further up the building, so that the cracks were largest on the top floor.

**Points for consideration**

– In structural systems where the ceiling slabs are supported both by the external and the internal walls, these load-bearing internal walls are exposed to a considerably greater load than the external walls as a result of dead and live loads. This additional load can be up to twice as heavy, depending on the direction of the span of the slab and the position of the internal walls. Moreover, if the support-forming central walls are of thinner construction than the external walls the stresses, and thus the elasticity in the internal walls, are considerably greater than those in the external walls.

– For reasons of sound and thermal insulation, different materials, with correspondingly different characteristics, are frequently used for external and internal walls.

– Creep ('plastic' deformation under long-term loads) in brickwork is proportional to the load and is also dependent on the characteristics (especially creep value, modulus of elasticity, brick crushing strength) of the building materials used. Similarly, there can be considerable differences in the shrinkage characteristics of the materials. Thus, the combined expansion resulting from creep and shrinkage of sand-lime bricks on the one hand and lightweight concrete and aerated concrete blocks on the other hand, is two and three times greater, respectively, than that of clay brickwork, even though all the materials are of the same strength. If, therefore, the materials used for the internal walls have higher creep and shrinkage values than those of the external walls, further differential expansion will occur between the two parts of the structure.

– These different types of deformation can be further amplified by temperature differences between the outside and the inside and by differences in the settlement of the foundations (see A 1.2.3).

– As the number of storeys increases, the bracing partition walls which are connected both to the external and internal wall are exposed to compounded differences in expansion. Experience has shown that from approximately the 4th or 5th storey, these differences become so great that they exceed the shearing and tensile strength of the brickwork and thus result in cracking in the vicinity of the external walls because the latter prevent any deformation. There is an increased risk of this type of cracking in thin walls (high height-to-length ratio) or if the brickwork is laid with insufficient bond.

Internal walls
Load-bearing internal walls

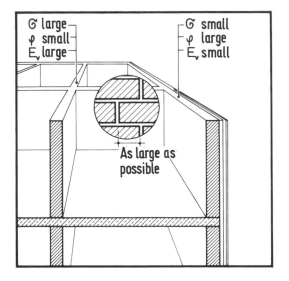

– Various criteria and formulae can be applied to calculate deformation, differential expansion or angular displacement and in this way it is possible to estimate, at least approximately, how free a structure will be from cracking. Your attention is drawn to the available literature on this subject.

**Recommendations for the avoidance of defects**

● In external and internal walls made of brickwork, the differential deformation caused by load and moisture should be as slight as possible. Moreover, in order to reduce differential expansion, bricks of a higher crushing strength than the structural calculations demand can be used for the internal walls, and bricks of a lower crushing strength for the external walls. Manufacturers should provide precise details of the shrinkage values of individual building materials.

Temperature differences between the internal and external wall should be reduced as far as possible (e.g. by means of double skin external walls or additional thermal insulation to the outside of the external wall).

● In load-bearing partition walls, the height-to-length ratio should, wherever possible, not exceed approx. 1:4 in 5-storey buildings, and approx. 1:3 in 3-storey buildings.

● The type of bond and the size of bricks used in the brickwork should be selected to ensure sufficient overlapping of the bond.

The bricks should be as dry as possible when supplied and laid and should be protected against becoming soaked once they have been laid.

In buildings with more than 4 storeys in particular, external and internal walls should be constructed of the same materials unless an approximate estimate of the risk of cracking has shown that the differential expansion will not have any harmful effect.

Internal walls
Load-bearing internal walls

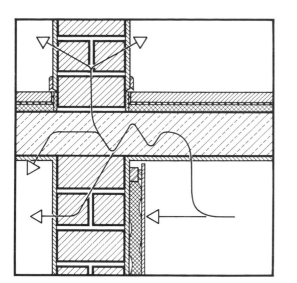

In some examples, load-bearing internal walls did not achieve the minimum levels of insulation against air-borne sound recommended. The walls concerned were single skin apartment dividing walls built of lightweight concrete bricks (gross density: 1.0 kg/m³) or walls between rooms with different sound level requirements (apartment/bar).

The insulation against air-borne sound provided by solid load-bearing walls inside apartments was also inadequate if the sound was able to penetrate adjacent rooms via open cavities between the wall and a timber joisted ceiling resting on it, or through flaws and open joints in non-plastered walls.

**Points for consideration**

– The acoustic damping provided by single skin load-bearing internal walls is dependent on their weight per unit area (density). However, it can be considerably impaired by incorrectly fitted additional coverings fitted to the surface of the walls, by indirect transmission and by flaws which continue through the whole of the wall's cross section. Thus, double skin wall constructions are necessary wherever the requirements in terms of acoustic insulation are high.

– Single skin brickwork walls which are plastered on both sides can achieve a maximum estimated sound insulation value of 55 dB (air-borne sound insulation = + 3 dB) with a weight per unit area of up to approx. 480 kg/m²; they are therefore not suitable for providing sufficient acoustic insulation as dividing walls for rooms with different sound levels (e.g. apartment/bar/music room) with a sound insulation value of 62 dB nor as the dividing walls between houses and apartments in terraced units (sound insulation value of 57 dB).

– If solid walls are clad with additional layers, this may worsen instead of improve sound insulation if rigid, sound proofing layers (e.g. lightweight woodwool panels, hard foam panels or similar) are fitted direct to the walls (resonance effect). Similarly, satisfactory sound insulation cannot generally be provided in this way, because indirect sound transmission (transmission through edges or linear sound conduction) cannot be sufficiently excluded. For the same reason, rigid soundproofing layers fitted in the core of external walls can result in a considerable deterioration in acoustic insulation between rooms.

– Flaws which continue through the whole cross section of a wall can cause a major reduction in sound insulation. Care should therefore be taken to ensure that all joints are fully sealed, especially in brickwork with continuous mortar filled butt joints. The risk of such continuous flaws can also be reduced by plastering both sides of the wall.

– If, for example in the case of timber joisted ceilings, the joists are supported by the wall, there are frequently voids in this area and thus increased sound transmission. Similarly, sound can also be transmitted to adjacent rooms through the air space above suspended ceilings if this cavity is not partitioned off or filled with sound absorbent materials, or if the suspended ceiling is covered with insulation layers which are either unsuitable (too rigid) or too thin.

Internal walls
Load-bearing internal walls

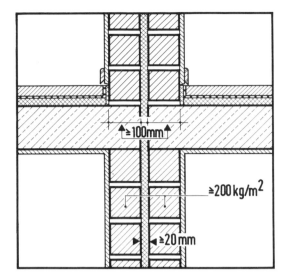

### Recommendations for the avoidance of defects

● Single skin brick walls provide sufficient sound insulation as partition walls between apartments (and within a single apartment) provided that they are of sufficient density ($\geq$ 480 kg/m²) and are adequately sealed (plastered on both sides) and provided that they have sufficient resistance to bending. For partition walls between houses (semi-detached or terraced houses) and for walls which separate apartments from rooms where there is a concentration of sound, only double skin walls can provide the necessary sound insulation.

● Double skin house partition walls must be separated by a continuous cavity across their width. This joint should, where possible, be 30 mm wide and should be filled with sound insulation materials which are as pliable as possible (e.g. mineral fibre insulation panels). Sound bridges caused by mortar droppings should be avoided in all instances.

● If, after they have been constructed, it is found necessary to improve the sound insulation characteristics of solid dense walls, additional skins which are as pliable as possible (e.g. plasterboard or lightweight woodwool panels) should be fixed by light framing to the front of the walls, providing the largest possible gap between the panel and the wall. This cavity should then be filled with soft insulation material (e.g. mineral wool). Indirect sound transmission should be avoided as far as possible by separating the framing members and the panels with sound absorbing pads (ceiling, side walls).

● There should be a 'soundproof' connection between the walls and the ceiling. In examples of suspended ceilings, the cavities should, where possible, be partitioned off, or should be filled with suitable insulation materials.

## Non-load-bearing internal walls

### Deformation

– Specialist books and guidelines

Grassnick, A.; Holzapfel, W.: Der schadenfreie Hochbau. Verlagsgesellschaft Rudolf Müller, Köln-Braunsfeld 1976.
Hartmann, M.: Taschenbuch Hochbauschäden und -fehler. Franck'sche Verlagsbuchhandlung, Stuttgart 1967.
Henn, W.: Die Trennwand. Callwey, München 1969.
Mayer, H.; Rüsch, H.: Bauschäden als Folge der Durchbiegung von Stahlbeton-Bauteilen. Deutscher Ausschuß für Stahlbeton. Heft 193, Verlag von Wilhelm Ernst und Sohn, Berlin 1967.
Meyer-Bohe, W.: Innenausbau. Verlagsanstalt Alexander Koch, Stuttgart 1976.
Reichel, W.: Ytong-Handbuch. Bauverlag, Wiesbaden und Berlin 1970.
Reichert, H.: Konstruktiver Mauerwerksbau, Bildkommentar zur DIN 1053. Verlagsgesellschaft Rudolf Müller, Köln-Braunsfeld 1976.
Rybicki, R.: Schäden und Mängel an Baukonstruktion. Werner-Verlag, Düsseldorf 1974.
Schneider, K. H. u. a.: KS-Mauerwerk. Herausgeber Kalksandstein-Information. Beton-Verlag, Düsseldorf 1975.
Weber, H.: Ausbauhandbuch. Karl Krämer Verlag, Stuttgart 1976.
DIN 1053, Teil 1: Mauerwerk; Berechnung und Ausführung. November 1974.
DIN 4103: Leichte Trennwände; Richtlinien für die Ausführung. Juni 1950.
DIN 4103 (Entwurf, Teil 1: Leichte Trennwände; Anforderungen, Arten. Dezember 1974.
DIN 18163 (Entwurf), Teil 2: Wandbauplatten aus Gips; Leichtwände, Richtlinien für die Ausführung. August 1974.
DIN 18330 – VOB, Teil C: Mauerarbeiten. September 1976.

### Specialist papers

Arnds, W.: Rißbildung in tragenden und nichttragenden Innenwänden und deren Vermeidung. In: Forum Fortbildung Bau 9, Forum-Verlag, Stuttgart 1978, Seite 109–116.
Arnds, W.; Vogt, J.: Leichte Trennwände auf weitgespannter Decke – Rißbildung in den Trennwänden. In: Bauschäden Sammlung, Band 2, Seite 98–99, Forum-Verlag, Stuttgart 1976.
Bundesverband Kalksandsteinindustrie e. V.: Nichttragende KS-Wände.
Casselmann, H.: Risse in gemauerten Trennwänden auf einer weitgespannten Stahlbetondecke. In: Baugewerbe, Heft 3/1977, Seite 22–23.
Pfefferkorn, W.: Nichttragende Innenwände unter massiver Dachdecke – Sind Risse in den Wänden immer Bauschäden? In: Bauschäden Sammlung, Band 2, Seite 100–101. Forum-Verlag, Stuttgart 1976.
Pieper, K.; Hage, D.: Risse aus Formänderungen des Mauerwerks. In: Ziegel 1971/72, Seite 90–99.
Pieper, K.: Risse im Mauerwerk, Teil I–III. In Deutsches Architektenblatt (DAB), Heft 2/1977, Seite 133–134; Heft 4/1977, Seite 295; Heft 6/1977, Seite 499–500.
Schild, E.: Leichte Trennwände auf Stahlbetondecke – Rissebildung in den Trennwänden. In: Bauschäden Sammlung, Band 3, Seite 88–91. Forum-Verlag, Stuttgart 1978.
Schubert, P.; Wesche, K.: Verformung und Rißsicherheit von Mauerwerk. In: Mauerwerk Kalender 1977, Seite 223–272.
Wesche, K.: Schubert, P.: Risse im Mauerwerk – Ursachen, Kriterien, Messungen. In: Forum Fortbildung Bau 6, Forum-Verlag, Stuttgart 1976, Seite 121–142.

## Sound insulation

– Specialist books and guidelines

Bobran, H. W.: Handbuch der Bauphysik. 3. Auflage, Vieweg, Braunschweig 1976.
Fasold, W.; Sonntag, E.: Bauphysikalische Entwurfslehre 4 – Bauakustik. Verlagsgesellschaft Rudolf Müller, Köln-Braunsfeld 1972.
Gösele, K.; Schüle, W.: Schall, Wärme, Feuchtigkeit. 3. Auflage, Bauverlag, Wiesbaden und Berlin 1976.
Schild, E.; Casselmann, H.; Dahmen, G.; Pohlenz, R.: Bauphysik – Planung und Anwendung. Vieweg, Braunschweig 1977.
DIN 1102: Holzwolle-Leichtbauplatten nach DIN 1101; Richtlinien für die Verarbeitung. April 1970.
DIN 1104 (Entwurf), Teil 2: Mehrschicht-Leichtbauplatten aus Schaumkunststoffen und Holzwolle, Richtlinien für die Verarbeitung. April 1978.

DIN 4109: Schallschutz im Hochbau. Teil 1: Begriffe. September 1962. Teil 2: Anforderungen. September 1962. Teil 3: Ausführungsbeispiele. September 1962. Teil 5: Erläuterungen. April 1963.
DIN 18165, Teil 1: Faserdämmstoffe für das Bauwesen; Dämmstoffe für die Wärmedämmung. Januar 1975.
DIN 18180 (Entwurf): Gipskartonplatten; Arten, Anforderungen, Prüfung. Juni 1977.
DIN 18181: Gipskartonplatten im Hochbau; Richtlinien für die Verarbeitung. Januar 1969.
DIN 68750: Holzfaserplatten; poröse und harte Holzfaserplatten; Gütebedingungen. April 1958.
DIN 68763: Spanplatten; Flachpreßplatten für das Bauwesen; Begriffe, Eigenschaften, Prüfung, Überwachung. September 1973.
DIN 18183 (Entwurf), Teil 1: Montagewände aus Gipskartonplatten; Richtlinien für die Ausführung. Januar 1975.

– Specialist papers

Gösele, K.: Schallschutz im Hochbau. In: Kalksandsteine – Bauphysik, Seite 10–31, Herausgeber: Kalksandstein-Information, Hannover 1977.
Steinert, J.: die Schalldämmung leichter zweischaliger Trennwände, In: Bundesbaublatt Heft 4/1970, Seite 163–169.
Volkart, K.: Die Luftschalldämmung von Wänden aus Gipskartonplatten. In: Das Baugewerbe, Heft 23/70, Seite 2040–2046.

## Load-bearing internal walls

### Deformation

– Specialist books and guidelines

Eggert, H.; Grote, J.; Kauschke, W.: Lager im Bauwesen. Verlag von Wilhelm Ernst und Sohn, Berlin 1974.
Frick/Knöll/Neumann: Baukonstruktionslehre, Teil 1. 25. Auflage, Teubner, Stuttgart 1975.
Funk, P.; Irmschler, H.-J.: Erläuterungen zu den Mauerwerksbestimmungen, Band I, DIN 1053, Blatt 5 und zugehörige Normen. Verlag von Wilhelm Ernst und Sohn, Berlin 1975.
Grassnick, A.; Holzapfel, W.: Der schadenfreie Hochbau. Verlagsgesellschaft Rudolf Müller, Köln-Braunsfeld 1976.
Hage, D.; Trautsch, W.: Formänderungen und Schertragverhalten von Mauerwerk. Berichte aus der Bauforschung, Heft 76. Verlag von Wilhelm Ernst und Sohn, Berlin 1972.
Hartmann, M.: Taschenbuch Hochbauschäden und -fehler. Franck'sche Verlagsbuchhandlung, Stuttgart 1967.
Reichert, H.: Konstruktiver Mauerwerksbau, Bildkommentar zur DIN 1053. Verlagsgesellschaft Rudolf Müller, Köln-Braunsfeld 1976.
Rybicki, R.: Schäden und Mängel an Baukonstruktionen. Werner-Verlag, Düsseldorf 1974.
Schild, E.; Oswald, R.; Rogier, D.; Schweikert, H.: Konstruktionsempfehlungen zur Altbaumodernisierung. Bauverlag, Wiesbaden und Berlin 1979.
Schild, E.; Oswald, R.; Rogier, D.; Schweikert, H.: Bauschäden im Wohnungsbau, Teil VI, Bauschäden an Kellern, Dränagen und Gründungen. Verlag für Wirtschaft und Verwaltung, Hubert Winge, Essen 1977.
Schneider, K. H. u. a.: KS-Mauerwerk. Herausgeber: Kalksandstein-Information. Beton-Verlag, Düsseldorf 1975.
DIN 1045: Beton- und Stahlbetonbau; Bemessung und Ausführung. Januar 1972.
DIN 1053, Teil 1: Mauerwerk; Berechnung und Ausführung. November 1974.
DIN 1054: Baugrund; zulässige Belastung des Baugrundes. Mit Beiblatt: Erläuterungen. November 1976.
DIN 4019, Teil 1: Baugrund, Setzungsberechnungen bei lotrechter, mittiger Belastung. September 1974.
DIN 4019, Teil 2: Setzungsberechnungen bei schräg und bei außermittig wirkender Belastung, Richtlinien. Februar 1961.
DIN 18330 – VOB, Teil C: Mauerarbeiten. September 1976.
DIN 18530 (Vornorm): Massive Deckenkonstruktionen für Dächer; Richtlinien für Planung und Ausführung. Dezember 1974.

– Specialist papers

Arnds, W.: Rißbildungen in tragenden und nicht tragenden Innenwänden und deren Vermeidung. In: Forum Fortbildung Bau 9, S. 109–116, Forum-Verlag, Stuttgart 1978.
Battermann, W.; Flohrer, M.: Lagerung massiver Flachdächer: In: Deutsche Bauzeitschrift (DBZ), Heft 11/1978, S. 1587–1593.

Internal walls
Typical cross section

Brandes, K.: Dächer mit massiven Deckenkonstruktionen. In: Berichte aus der Bauforschung, Heft 87, S. 5–37, Verlag von Wilhelm Ernst und Sohn, Berlin 1973.

Mann, W.; Müller, H.: Rißschäden bei Verwendung von Mauerwerk unterschiedlichen Verformungsverhaltens. In: Die Bautechnik, Heft 4/1975, S. 120–122.

Martens, P.: Risseschäden im Mauerwerk. In: Das Bauzentrum, Heft 6/1977, S. 5–11.

Meyer, H.-G. Verformung von Mauerwerk. In: baupraxis, Heft 5/1972, S. 53–59.

Pfefferkorn, W.: Dächer mit massiven Deckenkonstruktionen. In: Das Baugewerbe, Heft 18/1973, Seite 57–67; Heft 19/1973, Seite 54–59; Heft 20/1973, Seite 36–90; Heft 21/1973, Seite 54–63.

Pfefferkorn, W.: Längenänderungen von Mauerwerk und Stahlbeton infolge von Schwinden und Temperaturänderungen. In: Forum Fortbildung Bau 6, Seite 143–161, Forum Verlag, Stuttgart 1976.

Pieper, K.; Hage, D.: Risse aus Formänderungen des Mauerwerks. In: Jahrbuch Ziegel 1971/72, Seite 89–99.

Pieper, K.: Risse im Mauerwerk, Teil I–III. In: Deutsches Architektenblatt (DAB), Heft 2/1977, Seite 133–134; Heft 4/1977, Seite 295; Heft 6/1977, Seite 499–500.

Schubert, P.; Wesche, K.: Verformung und Rißsicherheit von Mauerwerk. In: Mauerwerk-Kalender, 1976, Seite 223–272.

Wesche, K.; Schubert, P.: Risse im Mauerwerk – Ursachen, Kriterien, Messungen. In: Forum Fortbildung Bau 6, Seite 121–142, Forum-Verlag, Stuttgart, 1976.

## Sound insulation

– Specialist books and guidelines

Bobran, H. W.: Handbuch der Bauphysik, 3. Auflage. Vieweg, Braunschweig 1976.

Fasold, W.; Sonntag, E.: Bauphysikalische Entwurfslehre 4 – Bauakustik. Verlagsgesellschaft Rudolf Müller, Köln-Braunsfeld 1972.

Gösele, K.; Schüle, W.: Schall, Wärme, Feuchtigkeit, 3. Auflage. Bauverlag, Wiesbaden und Berlin 1976.

Schild, E.; Casselmann, H.; Dahmen, G.; Pohlenz, R.: Bauphysik – Planung und Anwendung. Vieweg, Braunschweig 1977.

DIN 4109: Schallschutz im Hochbau. Teil 1: Begriffe. September 1962. Teil 2: Anforderungen. September 1962. Teil 3: Ausführungsbeispiele. September 1962. Teil 5: Erläuterungen. April 1963.

– *Specialist papers*

Brocher, E.: Die Verbesserung des Schallschutzes von Wohnungstrennwänden und Treppenraumwänden. In: Das Baugewerbe, Heft 23/70, Seite 2047–2057.

Gösele, K.: Die Luftschalldämmung von einschaligen Trennwänden und Decken. In: baupraxis. Heft 7/68, Seite 43–47.

Gösele, K.: Verbesserung des Schallschutzes im Wohnungsbau. In: Deutsches Architektenblatt, Heft 20/74, Seite 1379–1380.

Gösele, K.: Einige Probleme des Schallschutzes im Hochbau. In: Deutsches Architektenblatt, Heft 1/76, Seite 33–34.

Gösele, K.: Schallschutz im Hochbau. In: Kalksandsteine – Bauphysik, Seite 10–31. Herausgeber: Kalksandstein-Information. Hannover 1977.

Lutz, P.: Wohnungstrennwände aus Normalbeton, Mehrschicht-Leichtbauplatten und Beschichtung; ungenügender Luftschallschutz. In: Bauschäden Sammlung, Band 3, Seite 108–109, Forum-Verlag, Stuttgart 1978.

Lutz, P.: Einschalige Reihenhaus-Trennwand: Ungenügender Luftschallschutz infolge Schallängsübertragung durch Decke. In: Bauschäden Sammlung, Band 3, Seite 110–111, Forum-Verlag, Stuttgart 1978.

Pohlenz, R.: Mangelhafter Luftschallschutz zwischen drei Einfamilienreihenhäusern. In: Baugewerbe, Heft 24/77, S. 18–19.

Zimmermann, G.: Haustrennwand zwischen Reihenhäusern, ungenügender Luftschallschutz- und Körperschallschutz. In: Bauschäden Sammlung, Band 3, S. 106–107, Forum-Verlag, Stuttgart 1978.

## Problem: Plasters, claddings, tiling

The surface finishes of the walls on the inside of a building have a wide range of functions to fulfil. In addition to their main function of making the surface of the structure visually pleasing, they may also be responsible for providing protection against moisture in 'wet rooms', or for providing thermal insulation on the inside surfaces of external walls, for storing heat or moisture in order to balance conditions inside the room, or for providing sound insulation. Finally, particular demands may be placed on them in terms of wear-and-tear resistance and easy cleaning. However, judging by the results of the survey of structural failure, the task of preventing damage is not so multi-sided: it is limited largely to the prevention of cracking and other damage to surface finishes and to the avoidance of damp penetration damage in 'wet rooms'.

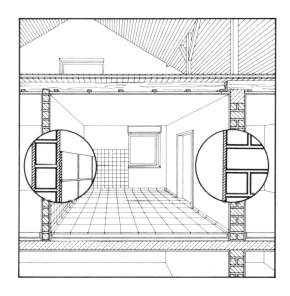

In terms of the materials and methods of application used for internal wallcoverings, there have been numerous changes over the last few years which have, in many cases, not followed Standard guidelines. This is, for example, true of the widespread use of single coat plasters and the almost exclusive bedding of internal wall tiles using the thin-bed method. These changes are reflected in the results of the representative building survey. Thus, in the following pages the main types of defect in wet plasters, claddings with plasterboard panels and chipboard panels and with tiling laid using the thin-bed method are discussed, and recommendations are made for the avoidance of defects.

Internal walls

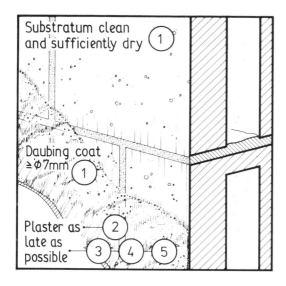

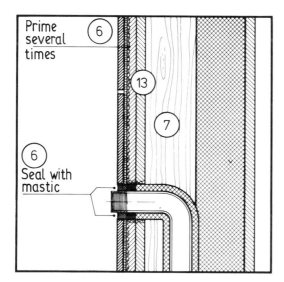

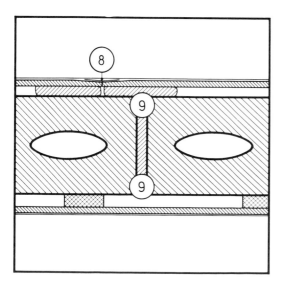

**1** The base for the plaster coat must be level and free of loose particles. Highly absorbent and smooth surfaces should be prepared with a daubing coat (addition of coarse sand 0–7 mm) which is suitable for both the plaster being used and for the substratum. The daubing coat should be applied at least 12 hours before plastering begins (see A 2.1.2).

**2** Where possible, plastering should not be carried out until several months after construction of the wall sections. If early plastering is unavoidable, care should be taken to ensure that the substratum is sufficiently dry (see A 2.1.2).

**3** If the internal wall finish takes the form of a two coat rendering 15 mm thick, the top and bottom coats of plaster should preferably be made of the same binding agent, with the strength curve running from the inside outwards (see A 2.1.3).

**4** Single coat adhesive renderings made of ready-mixed gypsum plasters should also be 15 mm thick. Care should be taken to ensure good room ventilation, especially when plastering in cold weather conditions. However, the fresh plaster should not be exposed to draughts or frost (see A 2.1.3).

**5** When selecting the material for the plaster coat, particular attention must be paid to ensuring that the plaster surface has sufficient strength for the intended wallcovering (see A.2.1.3).

**6** Gypsum plasters, gypsum plasterboard and chipboard must not be employed in wet rooms which get frequent use. Similarly, on wall surfaces in residential buildings which are exposed to spray water (showers), the wall materials, plaster and tile adhesives should consist of damp-proof materials. If, however, in such examples, the use of gypsum plasterboard or chipboard is unavoidable, then impregnated or semi-water-proof resin boards should be used; the cut edges of these boards should be primed with three coats of synthetic resin solution before installation and the surfaces should also be primed with three coats of synthetic resin after installation. Base layers of plaster should also be primed in the same way (see A 2.1.3, A 2.1.4, A 2.1.5, A 2.1.7).

**7** The size and spacing of the supporting framing and attachments and the method used for fixing gypsum plasterboard panels must follow the requirements of BS 5492: 1977 (see A 2.1.4).

**8** The edges of gypsum plasterboards should be jointed with synthetic resin filler compounds after inserting covering tapes (see A 2.1.4).

**9** Care should be taken to ensure that there are no open joints which continue through the wall's cross section (by taking measures such as plastering on one side, or careful jointing on both sides), especially when using gypsum plasterboard panels on external brickwork walls with claddings which are not airtight, and on brickwork walls which have to meet exacting requirements in terms of sound insulation (e.g. dividing walls between apartments) (see A 2.1.4).

**10** If it is necessary to apply a rendering coat in order to provide a good surface for laying tiles by the thin-bed method, this should take the form of a single coat at least 10 mm thick. Thin coats of fine plaster should be avoided (see A 2.1.6).

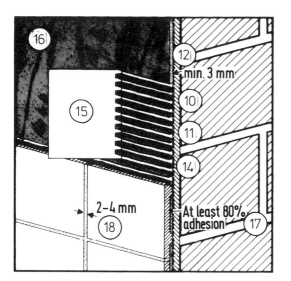

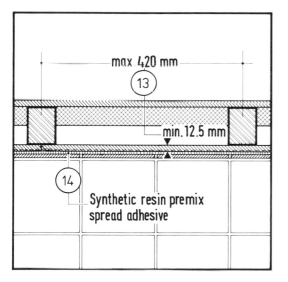

Synthetic resin premix
spread adhesive

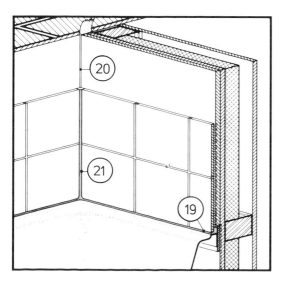

**11** In dry rooms or rooms with slight exposure to moisture (damp rooms in residential building), mortars from groups II or IV should be used as the preliminary rendering (see A 2.1.6).

**12** The surface of the rendering coat should be coarsely skimmed. Before priming with synthetic resin solutions, all loose particles, impurities and chalky deposits must be carefully removed from the surface of the rendering. The moisture content of the sub-stratum should be approximately 1 wt.% (see A 2.1.6 and A 2.1.8).

**13** If gypsum plasterboard panels are used as a basis for tiling, the panels should be at least 12.5 mm thick, and, with panels of this thickness, the spacing of the supporting framework should be no more than approximately 420 mm. However, if a supporting frame spacing of 625 mm is unavoidable, additional tie strips, or preferably two layer sheeting, should be fitted (see A 2.1.7).

**14** In the case of preliminary renderings and mortars from groups II and IV, and where tile laying surfaces are formed by chipboard or gypsum plasterboard panels, premix spread adhesives should be used in preference to hydraulically hardening thin-bed mortars for bonding tiles (see A 2.1.6, A 2.1.7 and A 2.1.8).

**15** The adhesive should be skimmed on with a float and then combed with a special comb to produce an even layer of adhesive at least 3 mm thick (see A 2.1.8).

**16** The area of wall covered with adhesive in one single operation should not be so large that the adhesive has started to cure before the tiles are laid (see A 2.1.8).

**17** The tiles must be pressed firmly into the bed of adhesive (they can be tapped in if necessary) in order to ensure that approx. 80% of the back of each tile is wetted (see A 2.1.8).

**18** As a guide, joints should be between 2 and 4 mm wide. The materials of which the grouting mortars are made should be of the lowest possible absorbency. When surfaces are exposed to moisture, epoxy resin grouting compounds are particularly suitable (see A 2.1.5, A 2.1.8).

**19** At junctions with adjacent structures and with materials with different deformation characteristics, as well as at connections with sanitary fittings, movement joints should be incorporated into the surface coverings. A movement joint should also be provided in the surface covering of structures which have to be separated for purposes of sound insulation (see A 2.1.9).

**20** Where anticipated movements are relatively small, and where cracking will result only in a less pleasing appearance, surface finishes should be separated from one another in such a way (e.g. trowel cut) that the expected cracking will take a regular form (see A 2.1.9).

**21** In examples of larger and repeated movements, the joint width should be chosen to match the linear changes which are expected. Joints which are exposed to water should be at least 5 mm wide and should be sealed with mastic jointing compounds after cleaning and preparation of the joint edges (see A 2.1.7, A 2.1.9).

Internal walls
Surface finishes

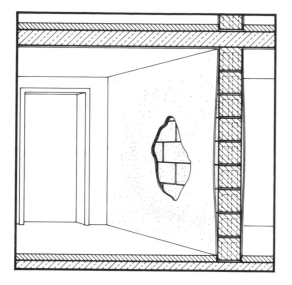

Inadequate adhesion of gypsum, lime or cement-lime internal plasters to walls frequently resulted in hollow areas and, in severe cases, in separation of the plaster skin over large areas of the substratum. In examples of damage covered by the survey, the substrata consisted of reinforced concrete, sand-lime bricks and lightweight concrete blocks, as well as wood-chip concrete blocks.

**Points for consideration**

– The absence of cracking in a plaster finish depends largely on good adhesion to the substratum.

– The adhesion of plaster to the substratum stems from the complex interaction of several factors. The mechanical keying of the plaster with the rough surface of the substratum when it is applied; keying as a result of the crystallisation of binding agents, which penetrate the capillary pores of the substratum, together with the mixing water; physical and chemical bonding forces between the materials of the plaster and the substratum. In freshly applied plasters, the adhesive force of the water and air pressure are largely responsible for the adhesion of the fresh mortar.

The importance of the individual processes mentioned above differs according to the type of mortar used (cement lime, lime or gypsum plasters, with or without admixtures to promote adhesion) and the type of substratum. In general, however, it can be said that the roughness, absorbency and moisture content of the substratum, as well as an even surface which is free of loose particles, are important factors in the adhesion of plaster.

– The survey of building defects revealed that of these various factors affecting internal wall plaster finishes, the moisture content and the absorbency of the substratum are the most important. If a wall is plastered too soon after it is built, the plaster will not, when applied, have sufficient adhesion to the substratum. Excessive wetting prior to application of the plaster has the same effect. If the absorbency of highly absorbent substrata is not reduced by means of a daubing coat, the water that the mortar requires in order to set may be removed and so the plaster coat is weakened in the area of the bond.

– Inadequate plaster adhesion does not generally become apparent until the adhesion area is exposed to shearing and tensile stresses. In internal walls, these stresses are generally the result of movements in the substratum. Contraction of the wall surface itself as a result of loading, or changes in moisture content, or deformation of wall surfaces that are supported on deflected ceiling slabs, can ultimately, if there is no joint along the top edge, result in direct compression forces in the wall plaster. However, since the majority of these movements occur in the first few months after the structures have been built, loads on the plaster finish can be reduced by delaying the plastering stage as long as possible.

Internal walls
Surface finishes

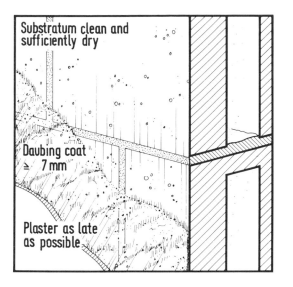

**Recommendations for the avoidance of defects**

● The base for the plaster coat must be level and free of loose particles. Highly absorbent and smooth surfaces should be prepared with a daubing coat (addition of coarse sand 0–7 mm) which is suitable for both the plaster being used and for the substratum. The daubing coat should be applied at least 12 hours before plastering begins.

● Where possible, plastering should not be carried out until several months after construction of the wall sections. If early plastering is unavoidable, care should be taken to ensure that the substratum is sufficiently dry.

Internal walls
Surface finishes

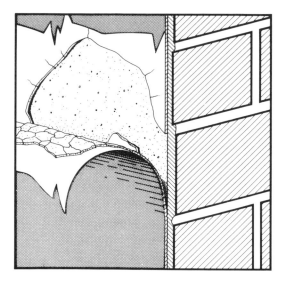

Thin, hard final coats of plaster (fine plaster) made of gypsum tended to pull away from the substratum when the latter consisted of mortars of lower strength, especially lime mortars from mortar group I. This defect was particularly noticeable when heavy wallpapers were applied to the wall or when wallpaper was changed during re-decoration. Damage to internal plaster in areas of the wall subjected to impact loads occurred if the plaster did not have sufficient strength – either because it was of the wrong composition, or because gypsum plasters were applied in wet rooms which were in frequent use.

**Points for consideration**

– In order to achieve a smooth finish on internal plaster, the final coat of plaster commonly takes the form of a layer of pure gypsum mortar a few millimetres thick. However, if this hard, brittle plaster skin is applied to a backing layer of plaster of lower strength which contains different binding agents, there may be poor adhesion of the final coat of plaster.

– Gypsum plasters swell slightly when they harden, unlike lime and cement lime plasters, which shrink. For this reason, and because of differences in the modulus of elasticity, stresses build up in the adhesion zone.

– Aerated lime mortars require carbon dioxide for the air to harden. Thus, if a dense final coat of plaster is applied too early, the backing layer of plaster will not be able to harden sufficiently and this will further reduce the strength of the bond.

– The final coat of plaster is particularly liable to peel away if coverings (such as hessian or flock wallpapers) are applied, which generate additional stresses.

– This risk of inadequate adhesion can be reduced if a period of several days is left between the application of the first coat of lime mortar and the gypsum plaster final coat, as well as by adding gypsum to the first coat of group I mortar. However, this damage can be more reliably avoided by adding the same binding agents to the first and final coats of plaster and by ensuring that there is a curve in the strength of the plaster from the inside to the outside. However, in doing this, the strength of the plaster surface must be suited to the designed surface covering – for example, group I plasters are not suitable as a basis for heavy wallpapers.

– Under no circumstances must gypsum be added to a backing plaster containing hydraulic binding agents (mortar groups II or III), since when they absorb water, ettringite crystals can form. These swell considerably and cause the plaster to peel.

– By using ready-mixed dry mortars to produce single coat adhesion plasters, the errors in mortar composition and in the sequence of coats can be avoided. However, the lime and cellulose admixtures in these ready-mixed mortars generally pre-set the surface of the plaster and produce a less absorbent surface. Moreover, if the mortar has to harden in rooms with high air humidity or where condensation forms on cold wall surfaces (when plastering in cold weather), hard layers of chalky deposits may form which will tend to part from the wall in a similar fashion to the thin layers of fine plaster described above.

– Gypsum mortar hardens by forming needle-like gypsum crystals which mesh in with one another and with the other plaster admixtures. Thus, if a gypsum plaster becomes wet for lengthy periods of time, it will lose its strength. For this reason, gypsum plaster is not suitable for wet rooms which are frequently used (e.g. public shower rooms).

Internal walls
Surface finishes

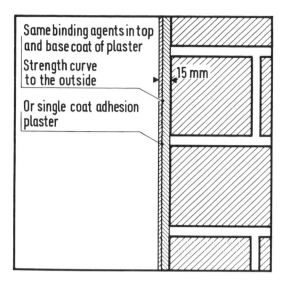

Same binding agents in top and base coat of plaster

Strength curve to the outside

Or single coat adhesion plaster

15 mm

### Recommendations for the avoidance of defects

● If the internal wall finish takes the form of a two coat rendering 15 mm thick, the top and bottom coats of plaster should preferably be made of the same binding agent, with the strength curve running from the inside outwards.

● Single coat adhesive renderings made of ready-mixed gypsum plasters should also be 15 mm thick. Care should be taken to ensure good room ventilation, especially when plastering in cold weather conditions. However, the fresh plaster should not be exposed to draughts or frost.

● When selecting the material for the plaster coat, particular attention must be paid to ensuring that the plaster surface has sufficient strength for the intended wallcovering.

● Gypsum plasters must not be employed in wet rooms that are in frequent use.

Internal walls
Surface finishes

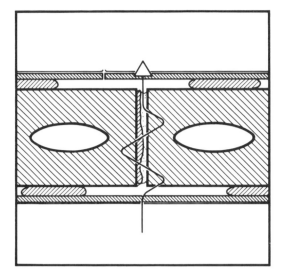

If inadequately water-proofed plasterboard panels were used in wet rooms, the panels warped and cracks were observed along the joints between them. Tiles became loose from plasterboard panels as a result of deformation of the backing if the spacing between the frame supporting the panels was too large. If no reinforcing tapes were fitted to the edges of the panels, these edges appeared as cracks in the plaster. If brickwork external walls were not fully jointed with mortar and were clad on the inside with plasterboard panels, or if the cavity behind the dry plasterboard skin was in contact with the ventilated cavity of flat roofs, harmful draughts occurred in the room at all connections which passed through the plasterboard wall (floors, sockets, pipe, conduits).

Apartment partition walls which were clad on both sides with plasterboard, and whose brickwork was also not fully jointed with mortar, provided insufficient sound insulation.

**Points for consideration**

– Gypsum materials swell when exposed for prolonged periods to heavy moisture and finally they lose their strength, since gypsum is soluble in water. Gypsum plasters and plasterboard panels are therefore not suitable for wet rooms with a large amount of water vapour (e.g. public baths).

– The water absorption of plasterboard panels can be reduced by impregnating and priming them. Factory impregnated plasterboard panels treated after installation with a synthetic resin primer can therefore be used in the presence of temporary slight exposure to water, such as in damp rooms in residential buildings. The plastic fillers used to seal the joints in the panels should also not swell or lose their strength when temporarily exposed to moisture and should therefore contain synthetic resins as binding agents.

– If the spacing between the supporting frames is too large, the panels will become deformed, and this will lead to loosening of rigid surface coverings (e.g. tiling) (see A 2.1.6).

– The varying movement of the edges of plasterboard panels resulting from deformation of the substratum and loading on the edges themselves can only be accommodated by conventional finishes and wallpapers if covering tapes are inserted over the edges of the panels.

– External walls must be airtight. In back-ventilated cavity wall structures, this function must be provided by the inner wall's cross section. When applying internal plaster, 'voids' in the brickwork of hollow-jointed walls or dry-laid blocks must be sealed on the inside. Plasterboard panels should, for ventilation purposes, be fitted to the wall with individual daubs of mortar, and since there must be a 2 mm wide joint at the floor and a 5 mm wide joint at the ceiling, there is formed behind the dry plaster surface a cavity which is connected to the inside of the room. Plasterboard claddings can therefore not be airtight.

– Since plasterboard panels applied with fixing mortar to both sides of the wall are unable to seal cavities (open joints) in the brickwork, the air-borne sound insulation of such walls is reduced, as these flaws have a great influence on sound transmission. Depending on the type of brickwork, this deterioration can amount to between 3 and 6 dB. If, however, one of the plasterboard panels is bonded on to mineral fibre tapes, the resultant double skin wall provides better sound insulation than a wall finished with wet plaster.

Internal walls
Surface finishes

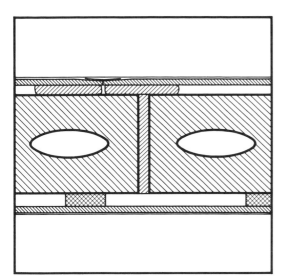

**Recommendations for the avoidance of defects**

● Gypsum plasterboard panels must not be employed in damp rooms which get frequent use. In damp rooms in residential buildings (kitchens, bathrooms) these panels should be impregnated and primed with synthetic resins.

● The edges of gypsum plasterboard panels should be jointed with synthetic resin filler compounds after inserting covering tapes.

● The size and spacing of the supporting framing and attachments and the method used for fixing gypsum plasterboard panels must follow the requirements of BS 5492: 1977 (plasterboard panels as a basis for tiling, see A 2.1.6).

● Care should be taken to ensure that there are no open joints which continue through the wall's cross section (by taking measures such as plastering on one side, or careful jointing on both sides), especially when using gypsum plasterboard panels on external walls with claddings which are not airtight, and on brickwork walls which have to meet exacting requirements in terms of sound insulation (e.g. dividing walls between apartments). If plasterboard panels are attached to brickwork walls with mineral fibre tapes, the air-borne sound insulation characteristics of the wall can be improved.

Internal walls
Surface finishes

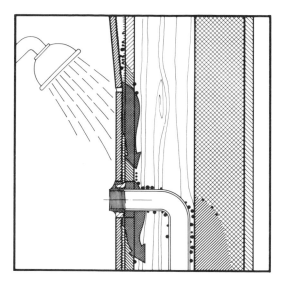

Damage occurred in internal wall surfaces in residential buildings which were exposed to water when water was able to get behind tiles which had been applied with gypsum mortar or with a spread adhesive. This resulted in loosening of the surface finishes.

## Points for consideration

– Compared with wet rooms in public baths (e.g. shower rooms) or in commercial premises, the exposure to moisture in wet rooms in residential buildings is slight. The costly sealing measures for wall surfaces can, therefore, generally be disregarded in residential buildings. However, brief exposure to spray water in residential buildings must be borne in mind, especially in areas near showers.

– The building surveys revealed that in examples of walls in residential buildings which are exposed to moisture, it is less a question of using sealing measures to prevent moisture from penetrating adjoining rooms than of preventing damage to the surface finishes that are exposed to water, and damage to the wall's cross section.

– Plastic and ceramic tiles are water-resistant and watertight. Water can therefore only get behind them through the joints in the tiling. However, once this occurs, it can result in damage which will finally lead to the tiles becoming detached from the wall. (See defect A 2.1.9, for sealing problems at wall junctions and connections with sanitary fittings.) Considerations at the design stage for the construction of damage-free wall surfaces in residential buildings must therefore aim at reducing or preventing the penetration of water through joints in tiling and at lessening the risk of damage by using substratum materials which are either protected against, or are insensitive to, moisture.

– Water can penetrate the tile covering through the jointing material itself and through hairline cracks at the edges of the tiles. The effects of even slight flaws can be immense, since the quantities of water which can enter by capillary action are much greater than the quantity of moisture which is lost by diffusion and evaporation.

– Ready-mixed jointing mortars for tile layings which are exposed to water (white cements) generally contain aggregates to reduce their absorbency, and modulus of elasticity, and to improve their adhesion. Investigations have, however, shown that the absence of hairline cracks at the edges of the tiles, and the absence of water penetration through the jointing material, can only be obtained with a high degree of reliability if epoxy resin-based grouting compounds are used.

– Gypsum materials, such as backing finishes made of gypsum plaster, plasterboard, or complete gypsum building panels, lose their strength if exposed to water for prolonged periods; moreover, if these materials are used in conjunction with hydraulic binding agents, swelling can occur. Similarly, chipboard panels are only water-proof to a limited extent. Many premixed adhesives swell when they absorb water. In principle, the risk of damage can be reduced if these materials are not used on walls which are exposed to spray water. Where the substratum is suitable (e.g. concrete), water-proof mortar from mortar group III should be used as the backing plaster, instead of gypsum mortar from mortar group IV, and cement bonding mortar should be used instead of premix spread adhesives.

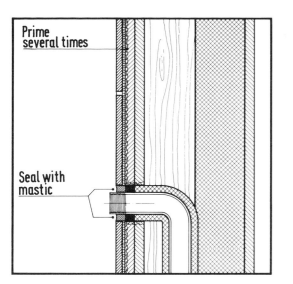

Prime
several times

Seal with
mastic

– However, in residential building, it is frequently not practical, either from a design point of view (e.g. in prefabricated houses consisting of chipboard-clad frame constructions, or in internal walls made of gypsum plasterboard panels), or from a working point of view (e.g. when plastering all other internal surfaces with gypsum plasters), to use water-insensitive substratum materials on the limited square metres of wall area which are exposed to spray water. In these examples, however, it is essential first to reduce the risk of damage as much as possible by choosing substrata which are as resistant to water as possible and by treating the surface of the substratum with water-repellent primers, and secondly to prevent the ingress of moisture by careful, high quality jointing, or by using joint-free surface coverings.

### Recommendations for the avoidance of defects

● In examples of wall surfaces in residential buildings that are exposed to spray water (showers), the wall materials, plasterboard surfaces, and tile adhesives should consist of damp-proof materials. If, however, the use of gypsum plasterboard or chipboard is unavoidable, then impregnated or semi-water-proof resin boards should be used; the cut edges of these boards should be primed with three coats of synthetic resin solution before installation, and the surfaces should also be primed with three coats of synthetic resin after installation. Base layers of plaster should also be primed in the same way.

● In examples of walls that are exposed to spray water, the jointing material used must be of the lowest possible absorbency and shrinkage. The use of epoxy resin jointing grouts or seamless surface coverings is particularly advisable where substrata have only limited resistance to moisture.

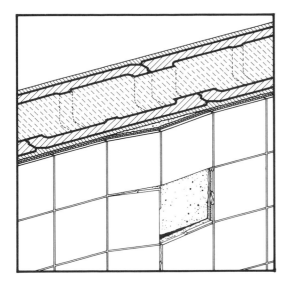

Tiles were particularly prone to become loose from the wall surface if they were applied to plasterboard wall surfaces using the thin bed method. These tiles came away together with their layer of adhesive, which was not bonded strongly enough to the substratum. In many examples a thin layer of fine plaster also came away – the weakest adhesion thus occurred between the top and second layers of plaster.

## Points for consideration

– Tiling applied using the thin bed method, in particular, requires good adhesion to the substratum, since, unlike tiles applied using a cement mortar bed, they are unable to support themselves or to accommodate flaws in the substratum.

– Adhesion to the substratum also depends on the strength of the substratum itself and on the bond between the adhesive and the surface of the substratum.

– Inadequacies in the strength of the substratum may above all be the result of:

– When using lime mortars, or excessively lean cement or gypsum mortars, the whole mortar coat is too weak. Moreover, in the case of aerated lime mortars, the early laying of tiles may have an adverse effect on the strength of the mortar, since the binding agent is unable to harden fully if it receives insufficient carbon dioxide from the air.

– Harmful variations in the strength of the plaster coat can also occur if it is applied thinly to even out irregularities in the substratum, thus causing large variations in its thickness. At particularly thin points, the water which the plaster coat requires in order to set, may be lost.

– Since the thin bed is unable to level out irregularities in the substratum, the preliminary coat of plaster is frequently covered with a thin coating of fine plaster in order to achieve an even surface. This layer of fine plaster generally contains more binding agent and is harder than the first coat. Indeed, in many examples, it is even made up of completely different mortar materials from the first coat. This results in poor adhesion between the first coat of plaster and the layer of fine mortar.

– If gypsum plasters become soaked as a result of a lack of priming, defective jointing of tile coverings and/or serious exposure of the wall to water, they lose their strength. The same applies if a gypsum plaster finish which has only dried on the surface is applied to damp brickwork (e.g. copiously wetted breeze blocks, fresh concrete) and is then covered with a watertight covering of tiles; in this instance, the plaster becomes soaked by the water in the substratum.

Defects in the bond between the adhesive and the substratum frequently occur under the following conditions:

– When loose particles, impurities, chalk deposits, or accumulations of binding agent on the surface (as a result of excessive rubbing down) prevent contact between the adhesive and the supporting substratum.

– In the absence of a synthetic resin primer to promote adhesion paste-type premix adhesive mortars do not wet the surface of the substratum sufficiently.

– Where cement bonding mortars are applied to gypsum plaster – and especially where moisture is able to reach the substratum because of the absence of priming – ettringite crystals can occur on the surface between the adhesive and the plaster.

Single layer backing plaster

Primer

Premix spread adhesive

– In the event of the presence of too much moisture (water-filled capillary pores) in the substratum, solvent bearing synthetic resin primers may not be able to penetrate deeply enough into the substratum.

– If the bonding mortar is applied direct with a comb, it is often not pressed firmly enough into the substratum.

– Inadequacies in the strength of the substratum and in the bond between the adhesive and the surface of the substratum have a particularly marked effect if the tiling forms a relatively rigid, high-stress skin on a soft substratum as a result of the use of cementitious bonding mortars. It is therefore always advisable to use the more flexible synthetic resin type of adhesive mortar, provided that the minimal moisture resistance of the latter will not prove a problem. However, if the strength of the substratum is similar to that of cement bonding mortars – e.g. a first coat of grade III cement mortar on concrete or sand-lime brickwork – tile adhesives which bond by hydraulic action are more suitable.

### Recommendations for the avoidance of defects

● If it is necessary to apply a rendering coat in order to provide a good surface for laying tiles by the thin-bed method, this should take the form of a single coat at least 10 mm thick. Thin coats of fine plaster should be avoided.

● In dry rooms, or rooms with slight exposure to moisture (damp rooms in residential buildings), mortars from groups II or IV should be used as the preliminary rendering. However, gypsum mortars from group IV should not be used in rooms that are heavily exposed to moisture.

● The surface of the rendering coat should be coarsely skimmed. It should be thoroughly primed with a synthetic resin solution. Before priming, all loose particles, impurities and chalky deposit must be carefully removed from the surface of the rendering. It is also essential to check – preferably with a moisture meter – that the moisture content of the rendering and of the adjacent brickwork surface is approximately 1–2 wt. %.

● If the tiled surface is to be exposed to only slight moisture for brief periods it is preferable to apply the tiles to group II and IV plasters with synthetic resin premix spread adhesives rather than cementitious bonding mortars.

### Mortar group chart for brickwork

| Mortar group | Non-hydraulic lime | Hydraulic lime | Cement | Sand |
|---|---|---|---|---|
| I | | 1 | | 3 |
| II | 2 | | 1 | 8 |
| IIa | 1 | | 1 | 6 |
| III | | | 1 | 4 |

### Mortar group chart for rendering

| Mortar group | Non-hydraulic lime | Hydraulic lime | Cement | Sand |
|---|---|---|---|---|
| Ib | | 1 | | 3 |
| II | 2 | | 1 | 9–11 |
| III | | | 1 | 3 |
| IV | 1 | | 1 | 3–4 |

Internal walls
Surface finishes

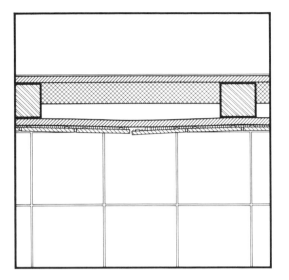

Ceramic wall tiles were frequently found to become loose from plasterboard or chipboard panels if they had been applied using the thin bed method. The supporting structure for the panels had warped, moved or, in instances of impact loads, had become deformed.

Cracking and loosening also occurred at wall connections and at junctions with sanitary fittings.

In many examples, the tiles were applied with cement bonding mortar. The damage here was particularly extensive if there was inadequate adhesion between the panels and the mortar.

**Points for consideration**

– Thin chipboard or plasterboard panels, and those fitted to widely spaced supporting frames, bend under impact loads.

– In lightweight framed walls with timber supporting studs, additional deformation can occur in the panelling as a result of warping of the frames if the latter were damp when installed.

– Ceramic tiles applied using the thin bed method are unable to accept this deformation. Thus, in order to prevent the tiles from loosening, the panels should be thicker and/or the spacing of the supporting frame should be less than is necessary and acceptable with other types of surface finishes. Double panelling has a particularly favourable effect.

– Chipboard and plasterboard panels lose their strength after prolonged and continuous exposure to water – chipboard is particularly liable to bulge if exposed to moisture on one side only. This also applies to resin-treated panels (treated to give limited weather-proofing) and factory impregnated gypsum plasterboard panels. Thus, when the surface is heavily exposed to water continuously (e.g. in shower rooms) these panels should not be used as a backing for tiles. Moreover, even where exposure to moisture is slight, for example in wet rooms (utility, kitchen or bathroom) in residential buildings, or where the tiles have been applied by the thin bed method and are largely water-tight, these backing materials can deform and weaken (especially at the edges of the panels and junctions for sanitary fittings and pipes) as a result of the penetration of water or condensation and can cause the tiles to become loose. However, if the surfaces of these panels, and particularly their edges and cut edges for openings, are primed with synthetic resin solutions before installation, and if pipe openings and wall connections are sealed with mastic silicon sealants before tiling, this risk of damage is reduced.

– Adhesives which set hydraulically (cement mortars) – even if organic admixtures are added to reduce their modulus of elasticity and improve adhesion – are relatively rigid and give lower adhesion values than synthetic resin premix adhesives. The latter, however, have less weather resistance. Nevertheless, since plasterboard and chipboard should only be used as backing materials for tiles where there is only slight and brief exposure to moisture, and since, therefore, the adhesive is not expected to be exposed to water, premix adhesives should be used in preference on these backing materials.

(For further details on tile laying and grouting see Defect, A 2.1.8).

Internal walls
Surface finishes

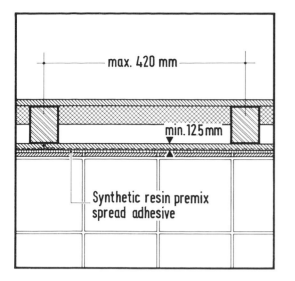

max. 420 mm

min. 125 mm

Synthetic resin premix
spread adhesive

**Recommendations for the avoidance of defects**

● Chipboard or plasterboard panels should be used as the substratum for ceramic tiles applied by the thin bed method only in dry rooms or rooms that have slight exposure to moisture for brief periods.

● With plasterboard, the panels should be at least 125 mm thick, and, with panels of this thickness, the spacing of the supporting structure should be no more than approximately 420 mm. However, if a supporting frame spacing of 625 mm is unavoidable additional tie strips or preferably two layer sheeting should be fitted.

● In examples of lightweight framed walls with timber supporting studs care should be taken to ensure that the timber is sufficiently dry.

● Once the substratum has been primed with a synthetic resin solution the tiles should be fitted with a synthetic resin premix adhesive (for details of laying and grouting see A 2.1.8).

● Factory-impregnated plasterboard and chipboard panels (resin-treated) should be used on wall surfaces with slight, brief exposure to moisture. Before installation, the edges of these panels should be treated with a synthetic resin premix solution to reduce their absorbency. Before the tiles are applied, the whole wall surface and all the cut edges of openings must be treated in the same way with three coats. Pipe openings and wall and floor connections in the tiled area should be filled with a mastic silicon sealant (see A 2.1.9).

Internal walls
Surface finishes

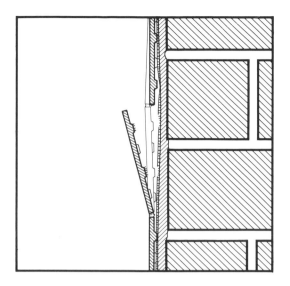

Defects in the bond, both between the adhesive and the substratum, and between the rear of the tile and the adhesive, caused the loosening of tiling which had been laid using the thin bed method.

Damp damage with subsequent tile loosening also occurred if water was able to penetrate a substratum made of gypsum materials through an uneven and cracked system of joints.

**Points for consideration**

– These stresses, which arise because of differences in the movement of tiling and substratum, are more easily accommodated by synthetic resin premix adhesives than by thin bed mortars, which set hydraulically and have a considerably higher modulus of elasticity. If cementitious mortars are used on gypsum materials, damp penetration can cause tiles to become loose because of the formation of ettringite. Most premix adhesives are less resistant to water than cementitious mortars. On substrata of relatively low strength, and on gypsum materials and panelled lightweight walls, premix ahesives are therefore better than cement mortars, provided that no wetting of the adhesive bed is anticipated.

– Thin-bed wall tiles are generally applied by the 'floating' method, in which the adhesive or mortar is applied to the substratum. However, if this is done directly with the comb, there is no guarantee of obtaining a strong bond between the adhesive and the substratum. However, if the adhesive compound is applied with a float, this will ensure that it is firmly pushed in to the surface.

– Very thin layers of adhesive are unable to smooth out even slight irregularities or cover sufficiently the back of the tiles – especially where their backs are grooved. The necessary thickness of the adhesive bed therefore depends on the grooves on the back of the tile. The layer of adhesive applied should be at least 3 mm thick.

– It is easier to achieve a uniform thickness by combing the layer of adhesive which has been applied. This also facilitates even tile laying.

– If a skin begins to form on the surface of the adhesive after it has been applied, this means that it is beginning to set, and strong bonding of the tile is no longer possible. The time which elapses between application and the commencement of setting depends on room temperature, room air-humidity, and the change of air inside the room. In examples of thin bed mortars which set hydraulically, skin can start to form as early as 10 minutes after application.

– If the tile is pressed home only lightly, this will not ensure that the whole of the back of the tile is covered with adhesive.

– Joints that are too narrow (less than 2 mm) are difficult to fill with pointing cement, while if the joints are too wide (more than 4–5 mm) there may be shrinkage cracks in the material.

– With conventional gypsum or cement-based pointing, shrinking and swelling movements in the pointing and deformation of the tiling itself and the supporting frame may give rise to hairline cracks at the edges of the pointing, which will let water through and which cannot be reliably avoided. Tiling pointed in this way is therefore not water-tight. However, by using epoxy resin grouting compounds, which, like other grouting materials, are applied in the form of a slurry by rubbing them into the joints with a sponge, this hairline cracking is largely avoidable and will help towards ensuring that the tiled surface is water-tight.

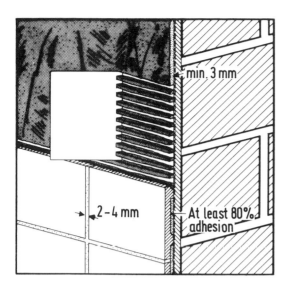

### Recommendations for the avoidance of defects

● Tiles should be applied only to a carefully prepared, level surface (see A 2.1.6 and A 2.1.7).

● In the case of preliminary renderings and mortars from groups II and IV, and where tile laying surfaces are formed by chipboard or gypsum plasterboard panels, premix spread adhesives should be used in preference to hydraulically hardening thin-bed mortars for bonding tiles.

● The adhesive should be skimmed on with a float and then combed with a special comb to produce an even layer of adhesive at least 30 mm thick.

● The tiles must be pressed firmly into the bed of adhesive (they can be tapped in if necessary) in order to ensure that 80% of the back of each tile is covered. This should be checked during laying by removing, at random, tiles which have just been laid.

● As a guide, joints should be between 2 and 4 mm wide. The materials of which the grouting mortars are made should be of the lowest possible absorbency. When surfaces are exposed to moisture, epoxy resin grouting compounds are particularly suitable. Most grouting compounds which are applied in the form of a slurry should be carefully rubbed into the joints with the aid of a grouting sponge.

Internal walls
Surface finishes

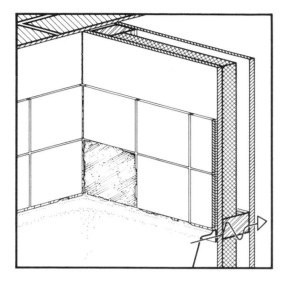

Cracking in surface finishes was frequently observed where these finishes connected to adjacent structures: e.g. at connections between ceiling and wall plaster, especially in examples of slab or timber joisted ceilings, and at connections between plasterboard-clad partition walls and plastered load-bearing walls. Cracks were also found where door frames were directly connected to plastered surfaces if these junctions were not covered.

In wall surfaces which were exposed to moisture, cracks in the corners of tiled partition walls, and particularly cracks at junctions with shower trays, baths and fittings, had serious consequences, since water was able to get behind the finishes and cause them to become loose.

In addition, the absence of movement joints at points where sanitary fittings were connected to walls resulted in increased transmission of sound through solid materials.

In all the examples covered by the survey the connections were plastered over or sealed without any special jointing methods.

In the area of metal plaster stops in expansion joints and corner stops in gypsum plaster, surface discoloration was noticed as a result of the rusting of the stops.

### Points for consideration

– At the junctions of varying surface finishes for different structures, changes in deformation occur which result in cracking, and sometimes even peeling, if the junction is covered with a material which does not readily follow deformation, e.g. plaster or gypsum filler.

– Whether, and by what means, this damage can be prevented depends mainly on the functions which the junctions have to fulfil. A distinction can be made between the following functions:

– The elongation, which is to be expected, is only slight and cracks will only mar the appearance of the work if they are irregular. A rebated joint will produce a problem-free rectilineal separation between the surface coverings. This can be achieved by cutting with a trowel, by inserting an adhesive strip before filling, or by using a groove. Typical applications are: at connections between wall and ceiling plaster in solid reinforced concrete slab ceilings; at connections between plasterboard walls and plastered walls.

These joints can, in general, finally be sealed when the building is redecorated for the first time, since the linear deformation results from shrinkage and creep deformation of the new building and should only occur once.

– The anticipated linear deformation is so large that open joint structures will spoil the appearance of the work. Moreover, linear deformation at the edges of the joints is recurrent. The width of the movement joint depends on the extent of the anticipated linear deformation. The joint must be provided with a covering or seal that will not be harmed by these movements. The following are suitable for making joints of this type: two-piece expansion joint metal plaster stops, covering strips which are attached to only one of the two adjacent structures, and slotted or grooved connecting rails. Typical applications are: the connecting joint at the support for freely mounted roof slabs, expansion joints, wall and ceiling connections of non-load-bearing internal walls, door frame linings.

Internal walls
Surface finishes

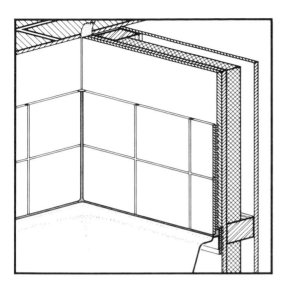

– The joint area is exposed to moisture. The joint must therefore be sealed with a mastic material. Elastic sealing compounds are generally used. The width of the joint should be selected in accordance with the expansion capacity of the sealant and with the linear deformation expected. However, joints should not be made narrower than 5 mm, since joints of this size cannot be filled correctly, and since in narrow joints it is very easy to exceed the expansion capacity of the jointing compounds. Moreover, the rear of the joint must be treated with hydrophobic primers, especially in the case of materials with only limited resistance to water. Typical applications are: wall connections of lightweight tiled partition walls where they are exposed to spray water, and connections for sanitary fittings.

– The joint should ensure that adjacent structures are separate from one another for the purposes of sound insulation. As a guide, the joint is sealed with a mastic sealing compound (as when exposed to moisture), or it is filled with soft material and covered. Typical applications are: connection of the surfaces of lightweight, double skin walls, as well as sanitary fittings.

– Gypsum materials allow ungalvanized sheet metal to corrode. Thus, only galvanized metal plaster stops or aluminium plaster stops should be used in gypsum mortar.

### Recommendations for the avoidance of defects

● At junctions with adjacent structures and with materials with different deformation characteristics, as well as at connections with sanitary fittings, movement joints should be incorporated into the surface coverings. A movement joint should also be provided in the surface covering of structures which have to be separated for the purposes of sound insulation.

● Where anticipated movements are relatively small, and where cracking will result only in a less pleasing appearance, surface finishes should be separated from one another in such a way (e.g. trowel cut) that the expected cracking will take a regular form.

● In examples of larger and repeated movements, the joint width should be chosen to match the linear deformation which is to be expected. Joints which are exposed to water should be at least 5 mm wide and should be sealed with mastic jointing compounds after careful cleaning, and preparation (priming) of the joint edges.

● Only galvanised metal plaster stops, or aluminium plaster stops, should be used in gypsum mortar.

Internal walls
Surface finishes

## Internal wall plasters

– Specialist books and guidelines

Albrecht, W.; Mannherz, U.: Zusatzmittel, Anstrichstoffe, Hilfsstoffe für Beton und Mörtel, 8. Auflage. Auflage, Bauverlag, Wiesbaden und Berlin 1968.

Arning, E.: Dispersionsfarben und Kunststoffputze. Verlagsgesellschaft Rudolf Müller, Köln-Braunsfeld 1974.

Brasholz, A.: Handbuch der Anstrich- und Beschichtungstechnik. Bauverlag, Wiesbaden und Berlin 1978.

Klopfer, H.: Anstrichschäden. Bauverlag, Wiesbaden und Berlin 1976.

Lade, K.; Winkler, A.: Ursachen der Putz- und Anstrichschäden. Verlag C. Maurer, Geislingen 1956.

Piepenburg, W.: Mörtel, Mauerwerk, Putz. 6. Auflage, Bauverlag, Wiesbaden und Berlin 1970.

DIN 18350 VOB Teil C: Putz- und Stuckarbeiten, August 1974.

DIN 18550: Putz, Baustoffe und Ausführung, Juni 1967, mit Beiblatt (Erläuterungen).

Beschichtung, Tapezier- und Klebearbeiten auf Innenputzen – Merkblatt Nr. 10 des Hauptverbands des deuschen Maler- und Lackierhandwerks, 1973.

– Specialist papers

Albrecht, W.: Über die Raumänderungen von Gips. In: Zement – Kalk – Gips, Heft 10, Seite 385, 1954.

Albrecht, W.; Wisotzky, Th.: Über die Härte von Innenputzen, in: bau + bauindustrie Heft 10 1968.

Grunau, E. B.: Neue Kunstharzdispersion als Zusatz für Mörtel und Putze. In: Baugewerbe, Heft 5, 1978.

Haagen, H.: Untersuchungen über Innenputze aus Werks-Trockenmörteln als Untergrund für Beschichtungen – Sonderdruck des Forschungsinstituts für Pigmente und Lacke e.V. Stuttgart, Anwendungstechnische Abteilung, 1978.

## Gypsum plasterboard panels

– Specialist books and guidelines

Hanusch, H.: Gipskartonplatten. Verlagsgesellschaft Rudolf Müller, Köln-Braunsfeld 1978.

Scheidemantel, H.: Handbuch der Plattentechnik. Verlagsgesellschaft Rudolf Müller, Köln-Braunsfeld 1969.

DIN 18181: Gipskartonplatten im Hochbau, Richtlinien für die Verarbeitung, Januar 1969.

– Specialist papers

Gösele, K.: Schalltechnische Eigenschaften von Trockenputz. In: IBP Mitteilung, Heft 1/1973.

Oswald, R.: Schäden an Oberflächenschichten von Innenbauteilen. In: Forum Fortbildung Bau 9, Referate Aachener Bausachverständigentage, 1978.

Poch, W.: Gipskartonplatte imprägniert – die Gipskartonplatte für den Feuchtbereich. In: Innenausbau Heft 3/1977.

## Tiling

– Specialist books and guidelines

Österreichisches Institut für Bauforschung: Abdichtungen und Abläufe bei Flachdächern, Dachterrassen, Balkonen, Loggien, Naßräumen, Wien 1973.

Pröpster, H.: Schadensanalysen bei Fliesen- und Plattenbelägen. Verlagsgesellschaft Rudolf Müller, Köln-Braunsfeld 1978.

Reichert, H.: Sperrschicht und Dichtschicht im Hochbau. Verlagsgesellschaft Rudolf Müller, Köln-Braunsfeld 1974.

DIN 4122: Abdichtung von Bauwerken gegen nichtdrückendes Oberflächenwasser und Sickerwasser mit bituminösen Stoffen, Metallbändern und Kunststoff-Folien, Richtlinien, Juli 1978.

DIN 18155 (Vornorm): Feinkeramische Fliesen, Teil 1–4, März 1976.

DIN 18156, Teil 1: Stoffe für keramische Bekleidungen im Dünnbettverfahren; Begriffe und Grundlagen, April 1977.

DIN 18156, Teil 2: Stoffe für keramische Bekleidungen im Dünnbettverfahren; Hydraulisch erhärtende Dünnbettmörtel, März 1978.

DIN 18352-VOB, Teil C: Fliesen- und Plattenarbeiten, August 1974.

Das Verlegen von Bodenfliesen und das Ansetzen von Wandfliesen; Merkblatt 1 der Fliesen-Beratungsstelle e.V., Großburgwedel 1969.

Die Fugen bei der Herstellung keramischer Wand- und Bodenbeläge; Merkblatt 4 der Fliesen-Beratungsstelle e.V., Großburgwedel 1969.

Ausführungsempfehlungen für den Einsatz keramischer Bekleidungen auf Leichtbauwänden aus Gipskarton und hölzernen Wandbauplatten; Fliesen-Beratungsstelle e.V., Großburgwedel, Sept. 1974.

Anwendungs- und verarbeitungstechnische Empfehlungen für die Verarbeitung von keramischen Fliesen auf Gipskarton- und Gipsbauplatten, Fliesen-Beratungsstelle e.V., Großburgwedel, Oktober 1973.

Richtlinien »Dünnbettmörtel für keramische Belagsmaterialien«, Fliesen-Beratungsstelle e.V., Großburgwedel, Dezember 1972.

– Specialist papers

Abfallschäden in Klebeverfahren, ihre Ursachen und Verhütung. in: Fliesen und Platten Nr. 15, 3. August 1970, Seite 253–256.

Hopp, E.: Abdichtung von Bauwerken gegen nichtdrückendes Oberflächen- und Sickerwasser; Richtlinien für die Ausführung, DIN 4122 (7/68) – Zusammenfassung der Bestimmungen im Hinblick für Abdichtungen unter keramischen Wand- und Bodenbekleidungen, Säurefliesner-Vereinigung e.V., Großburgwedel, Ausgabe II/1976.

Säurefliesner-Vereinigung e.V.: Ergebnisbericht zu Untersuchungen zur wasserabweisenden Eigenschaft von mit Kunstharzkitt auf Epoxidharzbasis ausgeführten keramischen Bekleidungen, Großburgwedel 1974.

Zimmermann, G.: Innenwände mit keramischen Wandfliesenbelägen für Duschräume; Rissbildung in Wänden, Abplatzungen der Fliesenbeläge, Bruch von Wasserleitungen. In: Bauschäden Sammlung Band 3, Seite 98–99, Forum-Verlag, Stuttgart 1978.

Zimmermann, G.: Keramische Wandfliesenbeläge auf Holzspanplatten-Wänden, Ablösungen der Fliesen von der Zement-Klebemörtel-Schicht. In: Bauschäden Sammlung Band 3, Seite 100–102, Forum-Verlag, Stuttgart 1978.

Zimmermann, G.: PVC-Wandfliesenbeläge auf Betonwänden, Ablösungen der Fliesen infolge Quellung und Verseifung des Klebers. In: Bauschäden Sammlung Band 3, Seite 101–105, Forum-Verlag, Stuttgart 1978.

# Problem: Reinforced concrete storey slabs

In residential buildings, the individual storeys are mainly formed by reinforced concrete, solid or ribbed slabs. Their primary structural role is to support loads resulting from dead loads (weight of the slab and the walls it supports) and imposed loads, consequently they are exposed to loads at right angles to their surfaces and are subject to deflection. This load-related deflection is composed partly of elasticity, which occurs immediately the load is applied, and partly by creep, which occurs over a longer period, depending on the characteristics of the concrete. Similarly, deflection is considerably increased by shrinkage processes which occur as the material dries out. It is only possible to limit the extent of this deflection by certain measures (e.g. weight imposed, span width, thickness of slab, type of concrete and the way in which it is laid and mixed). Such measures are all necessary if the slabs are to support walls in which there is a risk of cracking.

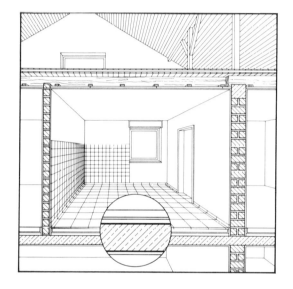

The majority of the damage related to reinforced concrete storey slabs was the result of deflection, although the consequences of this in the form of cracking were more commonly found in partition walls supported on the slabs, rather than in the slabs themselves. This deflection damage should therefore be regarded in close connection with damage to non-load-bearing or lightweight partition walls, especially since the damage often occurred even though the bending slenderness requirements had been maintained. These walls are particularly liable to suffer damage as a result of the long-term deformation which occurs after the wall is erected.

The above mentioned shrinkage and creep deformation frequently added to deformation of the other load-bearing structures (such as foundation settlement, differential expansion between external and internal walls) which thus changed the position of the supports and aggravated deflection; these influences are described in section A 1.2 – Load-bearing internal walls.

The following pages describe only the most common measures which have a direct effect on the deflection of reinforced concrete slabs – and in particular deflection which subsequently results in cracking in walls supported on these slabs.

Slabs

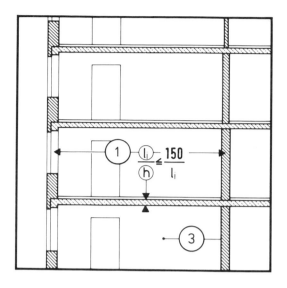

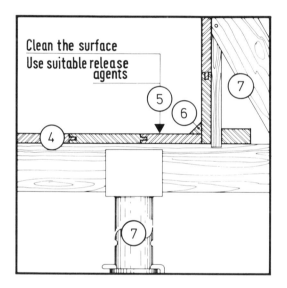

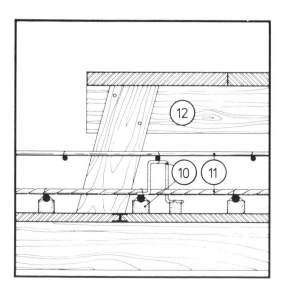

**1** The bending slenderness ratio of reinforced concrete slabs should be kept as small as is economically reasonable, especially in those with large span widths (irrespective of whether partition walls are to be erected on them or not) in order to ensure that floor or ceiling finishes applied to the slabs are level. The values obtained in the building should be below $li/h = 35$ or $li/h = 150/li$ (see B 1.1.2).

**2** If the bending slenderness requirements (of para. 1) cannot be met at reasonable cost by increasing the slab depth, or by reducing the spans between supports, possible cracks in the partition walls supported on the slabs must be overcome by constructional measures within the partition walls themselves, unless the cracks are accepted by the client (see B 1.1.2).

**3** In the case of wide-span reinforced concrete slabs and beams, additional measures should be taken to avoid cracks in partition walls (e.g. construction of self-supporting partition walls, sliding or flexible supporting frames, use of materials which will absorb deflection), even though requirements in terms of bending slenderness have been met (see B 1.1.2).

**4** The type of formwork used should suit the type of surface (exposed concrete surfaces to be plastered over). Formwork boards and shuttering should, as far as possible, be of the same thickness and should be fixed without voids. Timber formwork should be protected against long exposure to sun or wind and should be sufficiently dampened or oiled before the concrete is poured (see B 1.1.3).

**5** Before the concrete is poured, the formwork must be cleaned and treated with a releasing agent. The quantity, type and dilution of the releasing agent must be chosen in accordance with the material of which the formwork is made. Before the concrete is poured, the releasing agent must have dried properly and evenly (see B 1.1.3).

**6** All edges and corners should be chamfered by means of timber triangular fillets inserted into the formwork (see B 1.1.3).

**7** The structure supporting the formwork must be sufficiently strong to absorb the horizontal and vertical loads that occur. In particular, the formwork for the sides of walls and beams must be sufficiently supported to withstand the lateral pressure of the newly poured concrete (see B 1.1.3).

**8** The height of the formwork should be increased in accordance with anticipated and estimated deflection. It should not, however, exceed $li/300$ (see B 1.1.3).

**9** The striking times specified by Codes of Practice should be accurately observed. At low temperatures ($\leq + 5°C$) they should be extended. In order to minimise subsequent deflection as a result of creep, auxiliary supports should be erected immediately after striking and should be left in position as long as possible (see B 1.1.3).

**10** The reinforcement wires should be held in the necessary position by means of spacers. They must be secured to the reinforcing bars by tying or clamping so that they cannot move during the concreting process (see B 1.1.4).

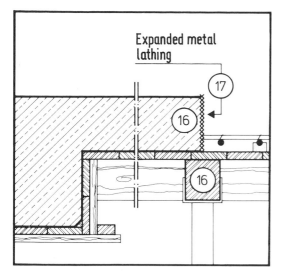

Expanded metal lathing

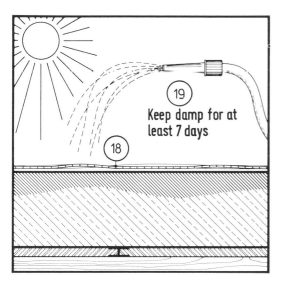

Keep damp for at least 7 days

**11** The diameter of the reinforcing wires should be such that there are sufficiently large voids between the individual bars to ensure that the concrete is able to surround them and become compacted in all the cavities. Before they are used, the reinforcing wires should be cleaned to remove any substances which might impede setting (see B 1.1.4).

**12** In order to protect the reinforcement from being trodden down during the concreting process, staging (catwalks) should be laid, with blocks on the formwork to prevent people walking on the reinforcement bars (see B 1.1.4).

**13** In examples of slabs that are supported on four sides, particular care should be taken to ensure correct positioning of the twisted reinforcing wires (see B 1.1.4).

**14** Careful checks should be carried out by supervisory staff during laying and concreting work to ensure that the reinforcement is correctly positioned and cannot move (see B 1.1.4).

**15** As soon as possible after mixing, the correctly poured concrete should be well compacted. With concrete of very soft consistency, attention should be paid to the risk of separation (see B 1.1.4).

**16** The concreting process should not be interrupted, but if the last layer of concrete to be poured has begun to set, joints must be provided or setting retarders used to extend the curing time. Joints in the work must be designed in advance, and must be made so that they do not occur in exposed areas of the formwork. If this is not possible, additional supports should be arranged directly beneath the area of the joint (see B 1.1.5).

**17** The connections between joints in the work must be made so as to ensure a dense and strong bond (see B 1.1.5).

**18** Until it has sufficiently hardened, the newly poured concrete must be protected against excessive cooling or heating, against drying out (as a result of the wind), against running water (rain), chemical attack, and vibrations and movements (see B 1.1.6).

**19** The concrete must be kept continuously damp for at least 7 days (14 days is better) by spraying or covering, e.g. with tarpaulins. Under certain conditions, spray-on wax-type treatment films may be suitable (see B 1.1.6).

**20** If there is a risk of frost, initial hardening can be speeded up by using rapid hardening cements. To protect them, the concreted structures (if necessary the whole site) should be temporarily covered (reinforced polythene sheeting) and the internal room heated. The newly poured concrete surfaces at least should be covered. In the event of longer periods of frost, the concrete can also be protected by correct heat treatment (steam, electric blower, infra-red heaters). In this case, formwork striking times – especially with solid concrete walls and slabs – should be extended (see B 1.1.6).

Slabs
Reinforced concrete storey slabs

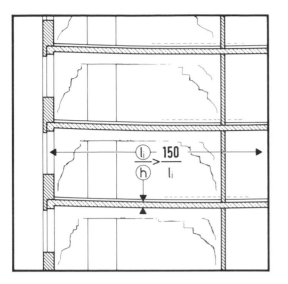

Reinforced concrete slabs, and beams with large spans and shallow depths in particular, became deflected to such an extent that in extreme cases the supporting structure itself cracked (at right angles to the supporting reinforcement), although the damage was generally limited to adjacent structures. Non-load-bearing internal walls supported on these slabs were especially susceptible to cracking, although there was also a tendency for plaster to spall from the supporting points of walls beneath these slabs, with a noticeable increase in sound transmission. This type of damage and defect occurred repeatedly, even though the structural components had acceptable bending slenderness.

**Points for consideration**

– When designing and selecting the size of slab, deflection is mainly influenced by the fixed depth, span and the rigid grid of the slab reinforcement. These three factors are known collectively by the term 'bending slenderness'. This describes the ratio of the span width to the depth of the slab, taking into account the type of support provided, since deflection is directly proportional to span width and indirectly proportional to slab depth.

– Moreover, deflection also depends on the amount of reinforcement used, since the use of additional steel reinforcement increases flexural resistance. Similarly, the use of compressive reinforcement can also reduce the influence of creep and thus, the deflection to be expected.

– Slabs with bending slenderness of specified limits (in the case of partition walls for load-bearing structures: $li/h \leq 150/li$) have suffered unacceptably large and harmful deflection, particularly if the subsequent, long-term deflection resulting from creep and shrinkage has been incorrectly estimated or calculated.

– Moreover, in the case of large span widths, the slab thicknesses become uneconomical if the necessary limit values for bending slenderness are observed. Thus, for partition walls supported on these slabs, it may be necessary to take other measures to avoid cracking (e.g. using timber stud framed walls, flexible preconstructed partitions (see A 1.1.4). Slight cracks which have no adverse effect on the strength of the partition wall can generally be permanently repaired once the major deflection deformation has occurred (after about 3 years, or the next time it is necessary to redecorate the building). If, however, the cracks can be accepted, the wall and storey slab can be constructed more economically. In any event, the developer must be informed of this 'risk of damage' by the architect or structural engineer.

– Deflected storey slabs can also apply pressure to partition walls beneath them. This will result in plaster spalling, and an increase in sound transmission because of the direct contact between the edges of the wall and storey slab. Moreover, deflection can also result in a limitation on use (furniture which will not be level, sticking cupboard doors) or will take the form of visible defects (sagging, uneven ceilings).

Slabs
Reinforced concrete storey slabs

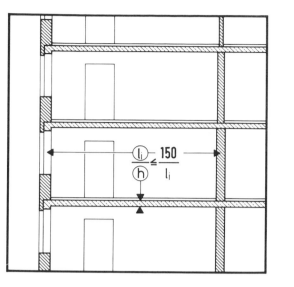

**Recommendations for the avoidance of defects**

● The bending slenderness ratio of reinforced concrete slabs should be kept as small as is economically reasonable, especially those with large span widths (irrespective of whether partition walls are to be erected on them or not), in order to ensure that floors or ceiling finishes applied to the slabs are level. The values obtained in the building should be below li/h = 35 or li/h = 150/li.

● If the bending slenderness requirements (of para. 1) cannot be met at reasonable cost by increasing the slab depth or by reducing the span widths between supports, possible cracks in the partition walls supported on the slabs must be overcome by constructional measures within the partition walls themselves, unless the developer is willing to accept the cracks.

● In examples of wide-span reinforced concrete slabs and beams, additional measures should be taken to avoid cracks in partition walls (e.g. construction of self-supporting partition walls, sliding or flexible supporting frames, use of materials which will absorb deflection), even though requirements in terms of bending slenderness have been met

Slabs
Reinforced concrete storey slabs

The most frequent types of damage and defect resulting from formwork errors were the following: pockets of unmixed gravel, damaged edges and corners, ridges, shuttering marks and irregularities on the undersides of slabs, bulging of the sides of beams, deflection of slabs and beams. The concrete did not provide a suitable surface for subsequent plastering, especially where it could not be repaired at a later stage, as in the case of mould oil residue or exposed concrete surfaces.

**Points for consideration**

– Highly dry timber formwork which is not dampened sufficiently before the concrete is poured will remove from the newly poured concrete parts of the water which it needs for complete hydration, and as a result, the concrete will not be able to harden properly – especially in slender structures.

– Formwork surfaces which have not been pre-treated are very difficult to strike. Incorrectly treated formwork surfaces (e.g. release agents applied unevenly, too thickly, or in too high a concentration, and especially where these are applied to formwork of poor and irregular absorbency), or the pouring of concrete into formwork on which the release agent (mould oil, wax or paste) has not dried sufficiently or consistently, will result in discoloration. Reduced strength and sanding will follow (especially if the aqueous emulsions of the mould oils mix with the moisture in the concrete, and thus to some extent penetrate the body of the concrete), or the surface will provide inadequate adhesion for the rendering coats to be applied to it (see B 2.1.3). Dirty formwork (wood shavings, sawdust, nails, paper or brick dust on the formwork) can reduce the amount of concrete cover over the reinforcement and may, as a result of this, lead to a reduction in load-bearing capacity.

– Deflected or loosely fitted shuttering of varying thickness, which allows fine mortar to escape through the joints, or timber formwork which subsequently swells, will produce ridges and uneven concrete surfaces which are an image of the shuttering.

– Sharp edges and corners are difficult to produce, since they are easily damaged during striking, or subsequently, as a result of impact. Moreover, coarse aggregate gathers in the corners (especially where there is insufficient compaction) and leads to pockets of gravel in these areas. All corners should therefore be chamfered by inserting timber triangular fillets into the formwork.

– The supporting structure for the formwork must not only support the weight of the newly poured concrete (if necessary with accumulations at various points), but must also be able to bear additional loads, depending on the way in which the concrete is delivered: size of delivery vehicle, drum tipping speed, and type of compaction. In addition, there are also horizontal loads (formwork pressure, wind loading, pressure from bracing struts, etc.) which must be dissipated by appropriate strengthening joints.

– Increasing the height of the formwork should prevent any irregularity where the slab connects with the support. This should therefore take into account the expected deflection in the ceiling. Increasing the height of formwork does not have any effect on subsequent deflection (especially as a result of shrinkage and creep) which can be harmful to partition walls standing on the deflected slab.

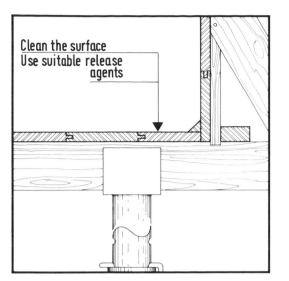

Clean the surface
Use suitable release
agents

– If the formwork is struck too early, with the concrete not sufficiently hardened – in other words with the concrete not yet able bear all loads with the required degree of safety – excessive stresses, severe plastic deflection and cracks may occur as a result of heavy shrinkage and creep. The striking times should be extended where there are low temperatures during the concrete hardening period. Moreover, extended striking times considerably reduce creep deformation of the concrete and deflection of slabs and beams, and retard drying speed, which has an important effect in limiting cracking.

**Recommendations for the avoidance of defects**

● The type of formwork used should suit the type of surface (exposed concrete, surface to be plastered over).

● Formwork boards and shuttering should, as far as possible, be of the same thickness and should be fixed without voids. Timber formwork should be protected against long exposure to sun or wind and should be sufficiently dampened or oiled before the concrete is poured.

● Before the concrete is poured, the formwork must be cleaned and treated with a releasing agent. The quantity, type and dilution of the releasing agent must be chosen in accordance with the material of which the formwork is made. Before the concrete is poured, the releasing agent must have dried properly and evenly.

● All edges and corners should be chamfered by means of timber triangular fillets inserted into the formwork.

● The structure supporting the formwork (girders, supports, bracing beams) must be sufficiently strong to absorb the horizontal and vertical loads that occur. In particular, the formwork for the sides of walls and beams must be sufficiently supported to withstand the lateral pressure of the newly poured concrete.

● The height of the formwork should be increased in accordance with anticipated and estimated deflection. It should not, however, exceed li/300.

● The striking times specified by Codes of Practice should be accurately observed. At low temperatures ($\leq + 5°C$) they should be extended. In order to minimise subsequent deflection as a result of creep, auxiliary supports should be erected immediately after striking and should be left in position as long as possible.

Slabs
Reinforced concrete storey slabs

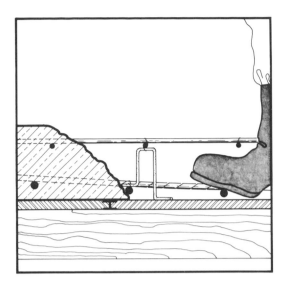

Reinforcement which had subsequently become displaced, or which had been incorrectly laid, was in some instances the cause of damage to reinforced concrete slabs and beams. The main types of damage found were cracks (e.g. on the surface of slabs supported on all four sides), chipping above the reinforcement, and signs of corrosion on the underside of the slab. In prefabricated reinforced concrete structures, damage such as chipping and corrosion was repeatedly found in the concrete parts, particularly at their junctions and in the area of supports.

**Points for consideration**

– If the positioning of the reinforcement is changed to such an extent that in the finished concrete structure it no longer corresponds to the reinforcement layout determined by the engineer's calculation, then the load-bearing capacity of the structure is questionable. Suitable spacers and staging should therefore be used to ensure that the reinforcement does not become displaced or trodden down. Hemispherical spacers made of concrete or mortar may be particularly unsuitable if they do not reliably permit void-free compaction, especially in the case of concrete of stiff or soft consistency.

– It is very difficult to return to their original position reinforcement bars which have been trodden down and distorted. This is particularly true of the top layers of reinforcement, which are exposed to negative moments (near supports, beams or cantilever slabs). Simple lifting of the steel reinforcement in the formwork when the concrete is poured must be regarded as a completely unsuitable method because of the lack of sufficient control.

– Any change in the depth of the reinforcement can lead to insufficient concrete cover and may result in chipping or signs of corrosion above the reinforcement.

– Cracks can never be totally prevented by means of reinforcement; however, if suitable reinforcement is used, instead of a few wide cracks there will be a large number of small cracks. In dry conditions (storey slabs and beams) crack widths of up to 0.2 mm should be regarded as harmless.

– Dirt, grease, oil, ice and loose rust can weaken the bond between the concrete and the steel and may result in severing the anchoring, and in cracks.

– Reinforcement which becomes concentrated (particularly at points where beams cross with supports) cannot be completely surrounded with concrete, since the concrete cannot be injected and compacted. This will frequently result in flaws, which may result in reduced load-bearing capacity and will at least give rise to corrosion.

– The lack of twisted reinforcement, or the use of insufficiently long twisted reinforcement bars to absorb the stresses in the corners of slabs, which are supported on all four sides, will give rise to diagonal cracks in the tops of the slabs. This damage can also be expected if the steel reinforcement has become trodden down, for example in slabs with mesh reinforcement.

– Since the supports for prefabricated reinforced concrete structures are generally not very deep, even a slight shift, or careless laying of the reinforcement, can result in damage to edges or cracking in the area of the supports.

Slabs
Reinforced concrete storey slabs

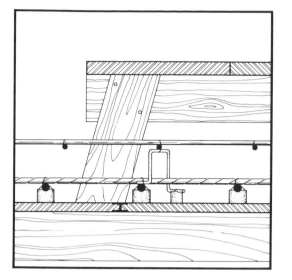

### Recommendations for the avoidance of defects

● The reinforcement wires should be held in the necessary position by means of suitable spacers. They must be secured to the reinforcing bars by tying or clamping so that they cannot move during the concreting process.

● The diameter of the reinforcing wires should be such that there are sufficiently large voids between the individual bars to ensure that the concrete is able to surround them and become compacted in all the cavities (approx. $1\frac{1}{2}$ times the diameter of the largest aggregate grain). If it is necessary to use internal vibrators (e.g. in relatively thick slabs and beams) suitable 'vibrator spaces' should be left in the reinforcement at intervals.

● In order to protect the reinforcement from being trodden down during the concreting process, staging (catwalks) should be laid with blocks on the formwork to prevent people from walking on the reinforcement.

● Any substances which might weaken the bond are to be removed from the reinforcement bars before they are used.

● In the case of slabs which are supported on four sides, particular care should be taken to ensure correct positioning of the twisted reinforcing bars.

● Careful checks should be carried out by supervisory staff during laying and concreting work to ensure that the reinforcement is correctly positioned and cannot move.

Slabs
Reinforced concrete storey slabs

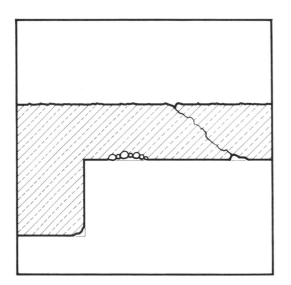

Even though the newly poured concrete had good properties and was correctly mixed, and despite correctly constructed formwork, there were repeated examples of the formation of pockets of gravel, and in severe instances, of defective concrete quality and/or reduced concrete strength. In these cases the concrete had been incorrectly introduced or worked. If joints in the work were not correctly made, pockets of gravel, uneven concrete surfaces and also cracking were the result. Complaints about these defects and faults were particularly common when the concrete surfaces were not plastered over, but remained exposed.

**Points for consideration**

– To obtain high grade concrete it is essential to produce a uniform texture which is as compacted as possible. Even as little as 8% vol. of additional pores can reduce the compression strength of the concrete by up to 50%, whilst excessive pore volume will also have an unfavourable effect on shrinkage and creep characteristics, since the increased volume of water in the pores will, once it evaporates, leave behind larger voids, which will result in a corespondingly larger reduction in volume.

– Incorrect pouring of the concrete into the formwork (e.g. with a conveyor belt without discharge hoppers, or excessive drops) results in separation (separation of the coarse aggregates from the fine mortar) which can be detected by the appearance of watery cement lime (laitance) on the surface.

– Inadequate compaction, especially of cements of stiff and plastic consistency, will not produce a homogeneous concrete slab in which the reinforcement rods are completely surrounded by concrete. However, in examples of very soft consistencies, excessive compaction can lead to separation of the concrete. The grading of the aggregate also influences the effect of compaction. Medium aggregate (e.g. 2–8 mm) may impede compaction since it can become lodged between the coarse aggregate and prevent the fine mortar from filling the voids. Moreover, without adequate compaction, it is particularly difficult for the concrete to get into the corners of the formwork, so that flaws and unmixed aggregates must be expected in these areas.

– Despite normal compaction, voids may form in densely reinforced or deep slabs beneath horizontal reinforcement rods, and beneath recesses, as a result of the concrete setting, and because of the accumulation of water beneath coarser aggregates (the cement 'leaches'). The concrete should therefore be compacted a second time in a way which is suited to the consistency of the concrete; this can increase its strength by up to 20%. The following tools are suitable for compaction and secondary compaction: manual or mechanical compactors (in stiff mixes); tampers with steel rods or wooden slats (in soft mixes); internal vibrators, surface vibrators or formwork vibrators (normal mixes).

– Secondary compaction may shake loose the reinforcement as a result of vibration in the concrete as it sets, and so the necessary adhesion will be lost. Secondary compaction should therefore not be carried out long after mixing (max. 7 hours), when the concrete has begun to set.

– Any joints that have to be made in the concrete (when work is interrupted, as a result of the weather, or because of mechanical breakdowns) must be able to absorb, without becoming damaged, any loads which occur. Impurities, cement slurry, concrete which has already hardened at the joint, as well as angled joint surfaces that are too smooth,

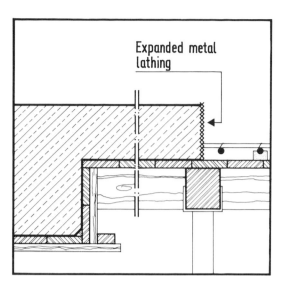

Expanded metal lathing

will not give adequate adhesion between the old and new concrete. Concrete which is too dry will remove from the newly poured concrete part of the water it requires to set, or will result in large differences in the shrinkage of the layers of concrete and may lead to cracks. Similarly, an excessive temperature gradient between the old and the new concrete will lead to stresses and possibly to cracks.

– If the joint in the work is made in the open part of the formwork, without direct support, differences in the deflection of the old concrete, which has already partially set, and the new concrete, may result in flaws on the underside of the slab, poor cohesion, and flaws in the consistency of the old concrete since it is unable to absorb the additional (shearing) stresses – particularly at the setting stage.

### Recommendations for the avoidance of defects

● The correctly poured concrete should be well compacted as soon as possible after mixing. The method of compaction used should suit the consistency of the concrete. Secondary compaction should be performed on particularly densely reinforced and deep slabs 3–3½ hours after mixing.

● With concrete of very soft consistency, the risk of separation should be borne in mind.

● The concreting process should not be interrupted, but if the last layer of concrete to be poured has begun to set (up to max. 24 hours, depending on the concrete mix), joints must be provided in the work, or setting retarders used to extend the curing time. Joints in the work should be designed in advance and must be made so as to ensure a dense and strong bond.

● Joints in the work should be struck vertically. Connections must be cleaned and, if necessary, slightly roughened. A strip of expanded metal inserted the full depth of the slab will promote a good bond. Old joints which have largely dried must be kept damp before further concrete is applied and should be warmed up at low temperatures in order to minimise shrinkage or temperature stresses.

● The joints in the work should be designed and arranged in such a way that they are not in an unsupported part of the formwork. If this is not possible, additional supports should be provided immediately beneath the area of the joint.

Slabs
Reinforced concrete storey slabs

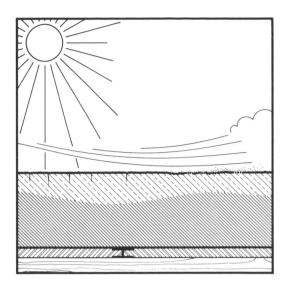

Cracking or exposed sandy deposits appeared in concrete slabs that had either not been treated, or had been incorrectly treated after construction and thus provided a poor surface for laying floor screeds. There was also evidence of loose textures and reduced strength, particularly in higher grade concretes.

## Points for consideration

- Despite correct composition, mixing, introduction and compaction, the concrete is unable to attain the required characteristics if it is not protected until it has hardened sufficiently. It is particularly important that the water it needs, to hydrate, is not removed prematurely, since a surface that dries too rapidly tends to suffer shrinkage cracks because of its low initial tensile strength and will have reduced bending strength because of the lack of water for complete hydration.

- In addition to the effect of excessive warming by direct sunlight, high air temperatures, and concrete temperatures, low relative air humidity and high wind speed increase evaporation and drying speeds; thus, damage caused by excessively fast drying out can also occur during winter months. When using wax-type covering films on the formwork to retard evaporation, care must be taken to ensure that this will not have an adverse effect on the adhesion of other layers to be applied subsequently to the concrete.

- Running water (heavy rain, excessive wetting of the laid concrete with a powerful jet of water) has an adverse effect on the water/cement ratio in newly compacted concrete and can flush out cement lime and fine components from the texture of the new concrete, resulting in permanent surface damage (sanding) which can reduce the load-bearing capacity, especially in the case of thin and ribbed slabs. Moreover, sudden cooling as a result of rain (and especially as a result of periodical hosing) can produce heavy and constantly changing temperature stresses, which can cause cracking in the new, still humid concrete. Excessive warming (direct sunlight), or cooling of the concrete have the same effects.

- Vibration and shaking of the concrete as it hardens (e.g. in the event of secondary compaction being performed long after setting, or if the formwork supports are not stable enough, so that it is exposed to vibration from passing traffic) can loosen the texture of the concrete and weaken the bond between the reinforcement and the concrete.

- The concrete is very sensitive to frost at the setting stage and when it begins to harden. At strengths of less than $5\,\text{MN}/\text{m}^2$ permanent damage will be done and strength lost if the concrete freezes even once. At low temperatures ($\leq 5°C$), hardening times are extended, and at less than $-10°C$ it is completely impossible for the concrete to harden and acquire strength.

Slabs
Reinforced concrete storey slabs

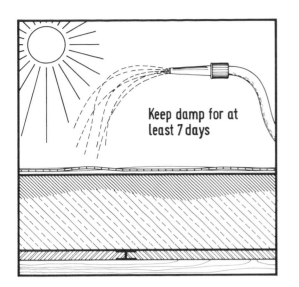

Keep damp for at least 7 days

### Recommendations for the avoidance of defects

● In order to ensure sufficient hydration, maintain the necessary temperature and moisture levels to achieve this, and limit early shrinkage, concrete with special properties such as water-proof, high frost or wear resistance, must be treated after it has been laid.

● Until it has hardened sufficiently, the newly poured concrete must be protected against excessive cooling or heating; against drying out (as a result of wind), against running water (rain), chemical attack, and vibrations and movements.

● Under normal conditions (15–25°C, 55–75% rel. air humidity) the concrete must be kept continuously damp for at least 7 days (14 days is better) by using sprinklers or perforated hoses, or by covering with tarpaulins. Under certain conditions, spray-on wax type treatment films may be suitable; these are sprayed on to the concrete after it has dried slightly, to delay evaporation by sealing the surface pores.

● If there is a risk of frost, initial hardening can be speeded up by using rapid hardening cements. To protect them, the concreted structures (if necessary the whole site) should be temporarily covered (reinforced polythene sheeting) and the internal room heated. The newly poured concrete surfaces, at least, should be covered with dry straw, or thermal insulation panels (where possible laid double or $\geq$ 30 mm thick). In the event of longer periods of frost, the concrete can also be protected by correct heat treatment (steam, electric blowers, infrared heaters). In this case, formwork striking times – especially with solid concrete walls and slabs – should be extended.

## Problem: Timber joist ceilings

Timber joist ceilings are multi-layer constructions, comprising a top layer (floor surface in storey slabs, sealing coats in the case of roof ceilings), load-bearing joists and a lower layer (ceiling for the storey below, made of plasterboard, lath and plaster etc.). In order to improve sound insulation, sand, pumice, or fine slag is laid on (and formerly between) the joists. Compared with solid concrete slabs, timber joist ceilings have the advantage of reduced weight, dry installation, and relatively high thermal insulation without the need for additional measures. On the other hand, however, they are sensitive to moisture and are not resistant to fungal or insect attack, or to fire, unless suitable protective measures are adopted. Their bracing effect is only slight because of their non-uniform internal structure, and since there is no compressed slab to distribute loads, individual loads can cause severe deflection and vibration.

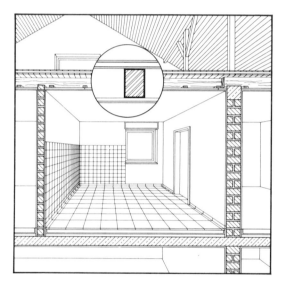

Today timber joist ceilings are used mainly in single or two-storey buildings as roof ceilings, and less commonly as storey slabs. The defects recorded with this type of structure were correspondingly few: defects concerning the cross section of timber joist ceilings could be placed in three categories – deformation, destruction by fungus, and defective sound insulation – which are described in further detail below.

The extensive defects found in double layer flat roof constructions with timber beams are discussed in detail in *Structural Failure in Residential Buildings, Volume 1: Flat roofs, roof terraces, balconies.*

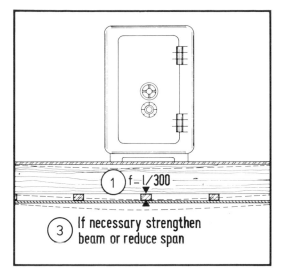

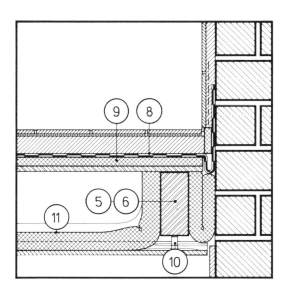

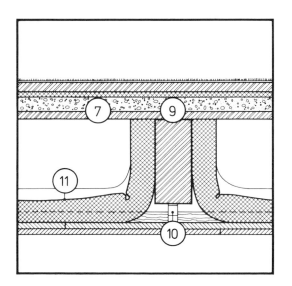

**1** The size of timber joist ceilings should be such that the loading placed upon them as a result of dead and live loads does not cause deflections in excess of span/300 for joists, and span/200 for purlins and rafters in pitched roofs (see B 1.2.2).

**2** The depth of the combined floor zone members should be increased by an amount equivalent to the deflection calculated from dead load and live load; other combination members should be increased in depth by at least span/300 (see B 1.2.2).

**3** If additional loads which were not anticipated at the design stage are to be imposed subsequently, the load-bearing capacity of the floor structure must be checked and if necessary they must be reinforced or supported to avoid deflection (see B 1.2.2).

**4** Where supporting beams carry secondary joists, changes in the shape and position of the central support in relation to the outer supports should be avoided as far as possible. The depth of the central support may have to be increased to correspond with the anticipated deflection (see B 1.2.3).

**5** According to Building Regulations, structural and chemical preservative measures must be taken to protect all timber members – especially those that are exposed to increased levels of moisture – and possible fungal attack. Before re-using old timber, it is imperative to make specific investigations for fungus, and if necessary preservative treatments must be applied (see B 1.2.3).

**6** If it is impossible to protect the timber from moisture during transportation, storage or erection, suitable arrangements must be made for the excess water to be removed in good time and without having any adverse effect on adjacent structures (see B 1.2.3).

**7** Filling material which is used to increase the weight of the structure should only be used if dry (see B 1.2.3).

**8** Timber members should be protected against rising damp or damp from adjacent structures by means of sealing courses (bond the seams). The penetration of moisture from bathrooms, etc. into the timber members must be prevented, for example by means of damp courses or vapour barriers. Failing this, moisture and rot-resistant materials should be used instead of timber (see B 1.2.3).

**9** The floor layer should be acoustically separated from the joists. This can be achieved by means of a sand filler layer laid over the joists, which will, at the same time, increase the floor dead weight. However, if the filling material is inserted between the joists, the upper floor layer should at least be laid on thin strips of insulating material (see B 1.2.4).

**10** Ceiling surfaces fitted beneath the floor or roof structure should be constructed as dense as possible in the form of a flexible layer (e.g. board shuttering with plaster, two layers of plasterboard) and – particularly if the upper layer cannot be separated from the joists – should be suspended flexibly and at individual points from the sound-transmitting supporting members. If this is not possible it should be attached to the joists with counter-battens; particular attention should be paid to obtaining close and flexible connections to the sides of all adjacent structures (see B 1.2.4).

**11** The ceiling void should be lined with sound-absorbent material (e.g. mineral fibre matting) (see B 1.2.4).

Ceilings
Timber joist ceilings

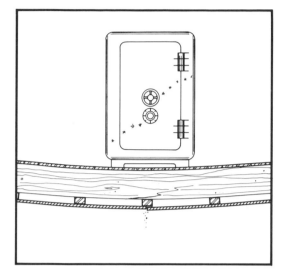

In several instances, considerable deflection was found in timber joists and this led predominantly to cracking, or parting, of ceiling finishes made of timber boards, plaster on laths, or plasterboard, or took the form of defects in the floor surface (sloping, non-alignment), or in the layer immediately beneath it.

The reasons for deflection in excess of the acceptable amount were mainly overloading of the timber joists as a result of changes in the use of the building, subsequent extension of roof spaces, the storage of heavy goods, subsequent gravel coatings or ponding on flat roofs, or the fact that the size of the timber joist was inadequate for the anticipated loading and span.

**Points for consideration**

- Like steel or reinforced concrete beams, timber joists deflect; this can be limited but not prevented. The amount of deflection which occurs is directly proportional to the load and span, and indirectly proportional to the moment of inertia and the modulus of elasticity of the joist.

- For timber box beams or similar construction used as ceiling joists, for example, in residential buildings, the maximum deflection resulting from dead load and live load is span/300 and span/200 for purlins and rafters.

- Increasing the depth of the floor zone members by between span/300 and span/200 has the effect of counteracting deflection, although the extent of this is dependent on the type of construction (box beams, ply beams), the way in which it is connected (bonded, bolted) and on the moisture content (unseasoned).

- Since for economic reasons the size of members is chosen on the basis of the loading known and anticipated at the design stage, the permissible deflection may subsequently be exceeded as a result of unforeseen additional loads (e.g. gravel layers or ponding on flat roofs), or additional loads applied at a later stage (e.g. the storage of heavy objects, subsequent extension of the roof space).

- Even where the deflection is large, the stability of the structure is not threatened, since the acceptable stresses (e.g. $= 10$ N/mm$^2$ in the case of grade II softwood, for example) are generally not exceeded; the deformation can, however, be transmitted to finishes fixed beneath the ceiling, where it can lead to damage, especially where joints around the edges are incorrectly made (see B 2.1).

- Where timber beams support secondary joists, increased deflection can also occur if the central supports are positioned lower than the outer supports, or if their position subsequently changes as a result of shrinkage, creep or deformation in the building site. As a result of settlement of the central support, the whole length of the beam deflects.

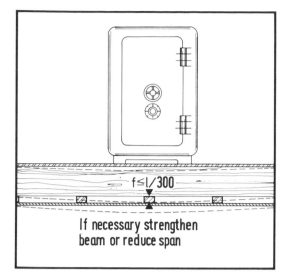

**If necessary strengthen beam or reduce span**

### Recommendations for the avoidance of defects

● The size of timber joist ceilings should be such that the loading placed upon them as a result of dead load and live load does not cause deflections in excess of span/300 for joists and span/200 for purlins and rafters in pitched roofs.

● The depth of the combined floor zone members should be increased by an amount equivalent to the deflection calculated from dead load and live load; other combination members should be increased in depth by at least span/300, or by span/200 if using semi-dry or unseasoned timber.

● If additional loads which were not anticipated at the design stage are to be imposed subsequently, the load-bearing capacity and possible deflection of the load-bearing structures must be checked and, if necessary, they must be reinforced or supported.

● Where supporting beams carry secondary joists, changes in the shape and position of the central support in relation to the outer supports should be avoided as far as possible. The depth of the central panel may have to be increased to correspond with the anticipated deflection.

Ceilings
Timber joist ceilings

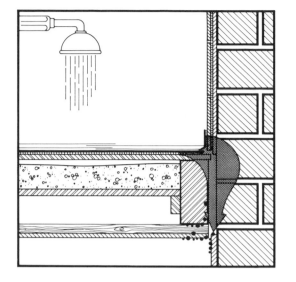

If timber joists were exposed to severe damp there occurred serious damage as a result of timber decay, caused mainly by dry and wet rot. Because of the latter, damp was able to penetrate the pores of the timber by various routes. Materials introduced when still damp such as concrete, timber supported on structures which were not damp-proofed (basement ceiling or floor slab in contact with the soil), penetration of spray water or water from plaster as a result of flaws in surface finishes or in seals between the floor and wall in bathrooms or utility rooms, or as a result of leaks in underfloor heating pipes, formed the main reasons for the appearance of fungal growth in building timber. Similarly, the timber joists in roof construction were affected by this type of damage where sealing layers were defective or ventilation inadequate.

## Points for consideration

– The optimum growth conditions for the fungus types which attack building timber – particularly the most dangerous type, which is dry rot – are temperatures of between 18°C and 22°C (room temperature), and a timber moisture content of about 28–30% (growth begins at about 20%). However, once growth has started, even dry timber can be affected.

– By breaking down the cellulose in the timber cells, dry rot results in the growth of destructive fungus; when this happens, the timber first turns a dark brown colour (brown rot) and then, at a later stage, cube-shaped pieces appear in the cracks; finally, the strength of the timber is completely lost when it becomes broken down into powder.

– Since in new buildings, dry rot is generally not spread by spores but by the use of timber which is already affected, there is a risk of dry rot anywhere where timber has not been preserved against fungal attack or when old timber is used in the building.

– During transportation, storage or erection, unprotected timber can become very damp as a result of rain or soil moisture. Moreover, timber which has already been erected can also be exposed to damp as a result of the prolonged presence of large quantities of moisture in the building, and particularly as a result of moisture being transmitted from adjacent structures or materials (e.g. external wall shuttering, timber supported on screeds or concrete which has not dried out sufficiently, or on floor slabs which are in contact with the soil, damp filling materials between beams) or because the timber is 'packed in' in largely air- and vapour-tight layers.

– Leaks in floorcoverings or wallcoverings, the lack of adequate sealing measures, or continuously high relative air humidity in damp rooms (kitchen, bathroom), as well as leaks in underfloor heating installations allow large quantities of moisture to enter the timber structure continuously.

– If timber structures are laid directly on to newly laid concrete slabs, they will not only be exposed to moisture resulting from the building process (surplus mixing water diffused from the concrete) but will also – in the case of basement ceilings or floor slabs in contact with the soil – be exposed to condensation, or moisture rising from the soil by capillary action. Here, the exposure to moisture is particularly severe if the floor does not have sufficient thermal insulation, or if it is separated from the inside of the room by a vapour resisting barrier (e.g. by a PVC floorcovering). (For further details see: *Structural Failure in Residential Buildings – Vol. III: Basements and adjoining land drainage.*)

Ceilings
Timber joist ceilings

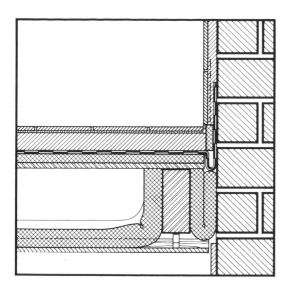

– In the roof space of double skin flat roofs, damage caused by condensation from wet rooms was common if the lower ceiling skin permitted vapour to pass through it or if it was not provided with sufficient thermal insulation, or if there was insufficient ventilation in the space. (For further details see: *Structural Failure in Residential Buildings – Vol. I: Flat roofs, roof terraces, balconies.*)

### Recommendations for the avoidance of defects

● According to Building Regulations, structural and chemical preservative measures must be taken to protect all timber members – especially those which are exposed to increased levels of moisture (those above or below wet rooms, on basement slabs, or floor slabs which are in contact with the soil) – against possible fungal attack. Before re-using old timber, it is imperative to make specific investigations for fungus, and if necessary, preservative treatments must be applied.

● If it is impossible to protect the timber from moisture during transportation, storage or erection, suitable arrangements must be made for excess water to be removed in good time, and without having any adverse effect on adjacent structures.

● Filling material laid on, or between, timber joists to increase density should only be used if dry.

● Timber members should be protected against rising damp, or damp from adjacent structures (e.g. from floor slabs which are in contact with the soil or from brickwork supports), or from adjacent building materials (e.g. damp filling materials) by means of jointless sealing strips (with bonded joints). The penetration of moisture from wet rooms into the timber members must be prevented by suitably reliable measures (jointless sealing courses or vapour barriers).

However, if this cannot be reliably guaranteed, timber construction should be dispensed with and a moisture- and rot-resistant structure used instead.

Ceilings
Timber joist ceilings

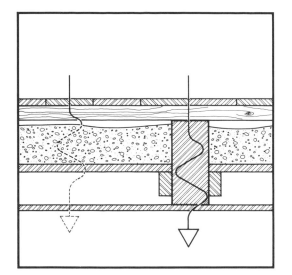

Storey ceilings in the form of timber joist constructions failed in some instances to satisfy minimum requirements in terms of insulation against air-borne and impact sound. The noise from a subsequently developed bar penetrated through the ceiling structure into the flat above to an unreasonable degree, despite the fitting (albeit incorrect) of an additional suspended ceiling. In this example, instead of an estimated sound insulation requirement of 62 dB (air-borne sound insulation = + 10 dB), the recorded sound insulation value was only 52 dB (air-borne sound insulation = ± 0 dB). In other examples, the floor finishes (chipboard) were directly fixed to the timber joists, so that it was impossible to obtain sufficient damping of impact noise. In addition, creaking noises were clearly audible in the flats below.

**Points for consideration**

– In timber joisted ceilings, the sound is mainly transmitted by the joists if the floor or the lower ceiling skin are directly fixed to the timber joists. Since the number of sound bridges cannot be reduced significantly because of the even spacing between the timber joists and since it is very expensive to construct separate supporting joists for the top and underside of the ceiling, the sound insulation characteristics of the structure depend on the way in which the top side and the underside are connected to the timber joists.

– The rigid, direct fixing of timber floorboards or chipboard (to act as a basis for floorcoverings) to the joists forms direct sound transmission surfaces. Moreover, these floors tend to creak under pedestrian use, especially if they are nailed.

– Direct firm fixings also have an adverse effect on the sound damping characteristics of the underside of the ceiling structure. The situation is made worse by, for example, the unfavourable sound reflection of light, rigid skins, or if faults in the surface or at the edges permit sound to pass through unhindered, or if the ceiling void is not filled with damping material, or is filled with unsuitable (not very porous) damping material.

– The otherwise favourable low weight of timber joisted ceilings has an adverse effect on sound insulation characteristics. Their weight should therefore be increased by means of filling material on or between the joists. This is, however, only effective if the upper skin is acoustically separated from the lower skin.

Ceilings
Timber joist ceilings

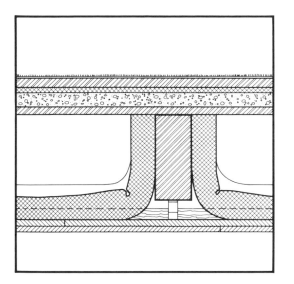

**Recommendations for the avoidance of defects**

● In timber joist ceilings, both the floor finish and the lower ceiling layer should be acoustically separate from the joists.

● The floor finish can be laid in a bed of sand above the joists, which will at the same time increase the density of the ceiling. If such a sand-filling layer is to be introduced between the joists, the upper floor layer should at least be laid on soft strips of insulating material placed on the joists.

Ceiling surfaces fitted beneath the floor or roof structure should be constructed as dense as possible, in the form of a flexible layer (e.g. board shuttering with plaster, two layers of plasterboard) and – particularly if the upper layer cannot be separated from the joists (e.g. with a sand filling) – should be suspended flexibly and at individual points from the sound-transmitting supporting members. If this is not possible, it should be attached to the joists with counter battens; particular attention should be paid to obtaining close and flexible connections to the sides of all adjacent structures.

● The ceiling void should be lined with a sound absorbent material which is as porous as possible (e.g. mineral fibre matting).

Ceilings
Typical cross section

### Reinforced concrete slabs

- Specialist books and guidelines

Grassnick, A.; Holzapfel, W.: Der schadenfreie Hochbau. Verlagsgesellschaft Rudolf Müller, Köln-Braunsfeld 1976.

Grün, W.: Beton 1. Werner Verlag, Düsseldorf 1969.

Hartmann, M.: Taschenbuch Hochbauschäden und -fehler. Franckh'sche Verlagsbuchhandlung, Stuttgart 1967.

Lohmeyer, G.: Stahlbetonbau für Techniker, Teil 1. B. G. Teubner, Stuttgart 1974.

Mayer, H.; Rüsch, H.: Bauschäden als Folge der Durchbiegung von Stahlbeton-Bauteilen. Deutscher Ausschuß für Stahlbeton, Heft 193. Verlag von Wilhelm Ernst und Sohn, Berlin 1967.

Rybicki, R.: Schäden und Mängel an Baukonstruktionen. Werner-Verlag, Düsseldorf 1974.

Weber, R.; Schwara, H.; Soller, G.: Guter Beton. Herausgeber: Bundesverband der Deutschen Zementindustrie e.V., Bauberatung Zement. Beton-Verlag, Düsseldorf 1976.

Wesche, K.: Baustoffe für tragende Bauteile, Band 2, Beton, Mauerwerk. Bauverlag, Wiesbaden und Berlin 1974.

DIN 1045: Beton- und Stahlbetonbau, Bemessung und Ausführung. Januar 1972.

DIN 18331 – VOB, Teil C: Beton- und Stahlbetonarbeiten. September 1976.

DIN 4235: Innenrüttler zum Verdichten von Beton; Richtlinien für die Verwendung. Oktober 1955.

DIN 4235 (Entwurf), Teil 1–5: Verdichten von Beton durch Rütteln. Januar 1976.

- Specialist papers

Arnds, W.: Materialgerechte Konstruktionen aus der Sicht des Sachverständigen. In: beton, Heft 5/1975, S. 161–165.

Bonzel, J.: Beton. In: Beton-Kalender, Teil 1, S. 5–91, Verlag von Wilhelm Ernst und Sohn, Berlin 1976.

Dartsch, B.: Vorbeugende Maßnahmen zum Vermeiden unerwünschter Risse im Beton. In: beton, Heft 4 4/1976, S. 130–134.

Grün, W.: Angewandte Betonprüfungen für Architekten. In: Deutsche Bauzeitschrift (DBZ), Heft 12/1972, S. 2471–2484.

Köneke, R.: Beton – problemloser Baustoff? In: Baugewerbe, Heft 11/1978, S. 31–34.

Ladner, M.: Schäden im Massivbau. In: Material und Technik. Heft 4/1975, S. 125–128.

Laengle, E. O.: Beschränkung der Durchbiegung von Stahlbetonbauteilen. In: Schweiz. Techn. Zeitschrift, Heft 8/1973, S. 137–147.

Mayer, H.: Verformungen von Stahlbetondecken und Wege zur Vermeidung von Bauschäden. In: Forum Fortbildung Bau 9, Forum-Verlag, Stuttgart 1978, S. 90–108.

Vollmer, H.: Bewehrungsarbeiten. Teilbericht zum Querschnittsbericht »Arbeitstechnik im Wohnungsbau«. In: Schriftenreihe 04 »Bau- und Wohnforschung« des Bundesministers für Raumordnung, Bauwesen und Städtebau, Heft Nr. 029, 1977.

Vollmer, H.: Befestigung der Bewehrung in der Schalung nach internationalen Regeln. In: baupraxis, Heft 2/1971, Seite 91–96.

### Timber joist ceilings

- Specialist books and guidelines

Bavendamm, W.: Der Hausschwamm und andere Bauholzpilze. Gustav Fischer Verlag, Stuttgart 1969.

Frick, Knöll, Neumann: Baukonsturktionslehre, Teil 1, 25. Auflage. Teubner, Stuttgart 1975.

Geiger, F.: Holzschutz. Werner-Verlag, Düsseldorf 1962.

Gösele, K.; Schüle, W.: Schall, Wärme, Feuchtigkeit. Bauverlag, Wiesbaden und Berlin 1977.

Gundermann, E.: Bautenschutz. Verlag Theodor Steinkopf, Dresden 1970.

Hartmann, M.: Taschenbuch Hochbauschäden und -fehler. Franckh'sche Verlagsbuchhandlung, Stuttgart 1967.

Rybicki, R.: Schäden und Mängel an Baukonstruktionen. Werner-Verlag, Düsseldorf 1974.

Schild, E.; Casselmann, H.; Dahmen, G.; Pohlenz, R.: Bauphysik – Planung und Anwendung. Vieweg, Braunschweig 1977.

Schmitt, H.: Hochbaukonstruktionen, 6. Auflage. Vieweg, Braunschweig 1977.

Scholz, W.: Baustoffkenntnisse, 8. Auflage. Werner-Verlag, Düsseldorf 1972.

Wahl, G. P.: Handbuch der Bautenschutztechniken. Deutsche Verlags-Anstalt, Stuttgart 1970.

Wesche, K.: Baustoffe für tragende Bauteile 4 – Holz, Kunststoffe. Bauverlag, Wiesbaden und Berlin 1973.

DIN 1052, Teil 1: Holzbauwerke; Berechnung und Ausführung. Oktober 1969.

DIN 4109, Teil 3: Schallschutz im Hochbau; Ausführungsbeispiele. September 1962. Din 18334 – VOB, Teil C: Zimmer- und Holzbauarbeiten. September 1976.

DIN 52175: Holzschutz; Begriffe, Grundlagen. Januar 1975.

DIN 68800: Holzschutz im Hochbau;
Teil 1: Allgemeines. Mai 1974.
Teil 2: Vorbeugende bauliche Maßnahmen, Mai 74.
Teil 3: Vorbeugender chemischer Schutz von Vollholz. Mai 1974.

- Specialist papers

Gösele, K.: Schalldämmende Holzbalkendecken. In: Informationsdienst Holz. Herausgeber: Entwicklungsgemeinschaft Holzbau (EGH) in der Deutschen Gesellschaft für Holzforschung (DGfH).

Graaff, Wille de: Bei Altbausanierung auf holzzerstörende Pilze achten! In: Baugewerbe, Heft 13/76, Seite 33–36.

Pfefferkorn, W.: Holzbalkendecke unter Dachgeschoß – zu große Deckendurchbiegung. In: Bauschäden Sammlung, Band 2, Seite 104–105. Forum-Verlag, Stuttgart 1976.

Tebbe, J.; Teetz, W.: Grundlagen zu Holz-Hochbau-Konstruktionen – Teil 2: Decken und Innenwände. In: Informationsdienst Holz – Wände, Decken und Dächer aus Holz – Konstruktion, Wärmeschutz, Schallschutz, Brandschutz, Seite 14–19, Herausgeber: Arbeitsgemeinschaft Holz e.V., Düsseldorf.

# Problem: Plaster finishes, claddings

In residential buildings, load-bearing ceiling structures are generally provided with surface finishes. Reinforced concrete slabs are generally plastered, and timber joist ceilings, which are still commonly used in newer houses for the pitched roof ceiling, are either plastered after a backing surface has been fitted to accept the plaster finish, or are clad with plasterboard or timber panels. Ceiling finishes can also play a special role in sound insulation (double layer ceiling structures), and room acoustics (the installation of sound-absorbing surfaces). Technical considerations relating to service installations (e.g. the laying of cables in the ceiling void) can also be a determining factor in the type of ceiling finish adopted. However, although the above-mentioned functions of ceiling surface finishes do not generally have to be fulfilled in residential buildings, the balancing effect of the surface finishes in terms of room conditions is important, both in terms of the storage and dissipation of heat and thermal insulation in the summer and in terms of the temporary storage of surface condensation in rooms of high humidity. However, the primary consideration in selecting a ceiling finish is that of achieving a surface which is as pleasing as it can be to the eye.

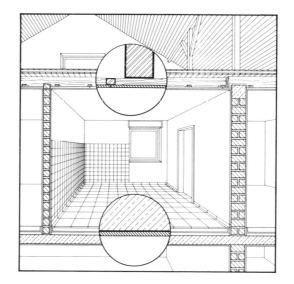

This is also the reason why the large volume of cracking, chipping, warping and deflection which the building survey revealed is regarded as a particularly important defect, even though the consequences were only relatively slight. However, in addition to this wide range of relatively minor damage, ceiling surfaces also suffered other forms of damage which take on particular significance, less because of their frequency than the risks associated with them: the spalling of surface finishes from a large area of a ceiling is one of the most serious examples of damage on the inside of a building, since it can cause serious injury to people in the room concerned.

In the following pages, the main types of defect in exposed concrete ceilings, plastered reinforced concrete, and timber joist ceilings, as well as plasterboard and timber layers are described and recommendations are made for the avoidance of structural failure.

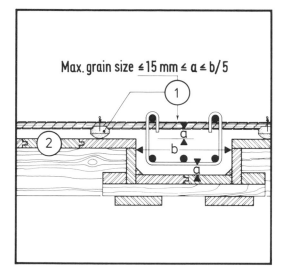

Max. grain size ≤ 15 mm ≤ a ≤ b/5

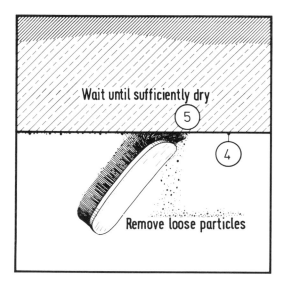

Wait until sufficiently dry

Remove loose particles

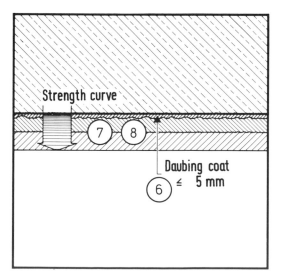

Strength curve

Daubing coat ≤ 5 mm

**1** In order to obtain perfect exposed concrete surfaces, particular attention should be paid to ensuring that the reinforcement is sufficiently covered. This is best done by fitting spacers. The maximum aggregate size ($\leq$ 15 mm) must be suited to the necessary concrete cover over the reinforcement, or to the smallest measurement of the slab or beam ($\leq$ 1/5) (see B 2.1.2).

**2** In order to avoid flaws in exposed concrete surfaces, the formwork boards must be laid as tightly as possible, they must be clean and must be treated with suitable release agents. In addition, the concrete must be well compacted (possibly compacted a second time) and must be given the correct after-treatment to suit the type of concrete and weather conditions (see B 2.1.2).

**3** In order to avoid discoloration and soiling, the concrete should contain only the same type of cement and uniform clean aggregates (see B 2.1.2).

**4** Before plastering reinforced concrete slabs, the condition of the plaster substratum must be carefully checked. In particular, the moisture content, absorbency, strength, and even surface of the concrete must be checked by close inspection, and by cleaning, scratching and wetting tests (see B 2.1.3).

**5** Where the substratum is still wet, wait until the concrete has dried further. Loose particles must be removed by brushing or sand blasting. Release agent residue should be removed by means of aqueous solvents (see B 2.1.3).

**6** In principle, ceiling plaster should be applied to a substratum which has already been treated with a daubing coat. The strength of the daubing coat should be at least equal to that of the coat of plaster to be applied to it. The aggregate used should be sand with a high proportion of coarse grains (diam. 5 mm), and it should be applied at least 12 hours before plastering begins (see B 2.1.4).

**7** Arrangements for plaster types and plastering methods – such as the use of adhesion promoting agents, or single coat adhesion plasters – should be agreed separately. Thin, single coat adhesion plasters should only be used on ceilings with fine surface tolerances and should not be less than 8 mm thick (see B 2.1.4).

**8** Gypsum plasters should not be used in rooms exposed to large amounts of moisture (see B 2.1.4).

Ceilings

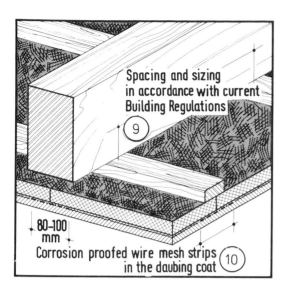

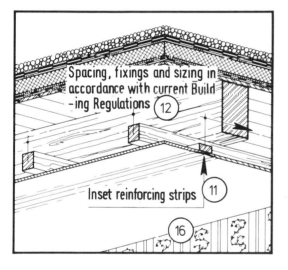

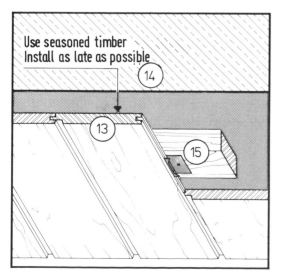

**9** When securing plaster supporting materials to the underside of the ceiling, the requirements of the relevant standards in terms of type, minimum dimensions, method of installation and corrosion resistance of the means of attachment must be taken into account. Careful checks should be made during construction to ensure that these requirements are being met (see B 2.1.5).

**10** When lightweight building panels are laid together and plastered dry, all the edges of the panels must be provided with corrosion-proofed wire mesh strips which are embedded in the daubing coat (see B 2.1.5).

**11** In examples of ceiling claddings made of plasterboard which are to be painted or papered, panels with bevelled, cardboard-covered edges should be used. All these edges should be filled after inserting 'reinforcing strips' (see B 2.1.6).

**12** When selecting the type and size of the attachments to be used, and the type and size of supporting structure, the provisions of the current Codes of Practice should be taken into account (see B 2.1.6).

**13** The moisture content of solid boards for ceiling claddings should be equivalent to 'balanced moisture' before they are installed. They must therefore at least be 'dry'; moreover, it is also advisable to allow them to adapt to the anticipated room conditions by storing them in the room for a reasonable period of time (see B 2.1.7).

**14** In new buildings, timber layers should be fitted beneath reinforced concrete ceilings as late as possible after the ceiling has been completed.

**15** The way in which the boards are secured must prevent them from warping, but should also impede expansion and contraction of the timber as little as possible. Indirect fitting by means of special brackets is a particularly good solution (see B 2.1.7).

**16** In principle, the ceiling plaster should be separated from adjacent wall surfaces by means of a joint. Where the anticipated movement is slight, it is sufficient if, before plastering, the ceiling and wall plaster are separated from one another, e.g. by means of a trowel cut or by using galvanized plastering stops. In the event of considerable movements (e.g. roof ceilings on flexible mountings) the width of the joint must be matched to the expected linear deformation. Subsequent filling with solid material should be prevented by installing two-piece plaster cove or covering beads (see B 2.1.8).

Ceilings
Surface coverings

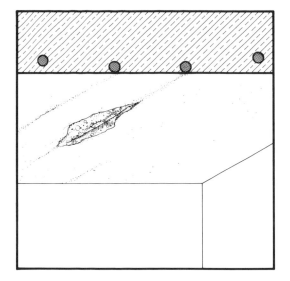

In keeping with the rare use of exposed concrete surfaces inside residential buildings, the defects and faults found in this case were not very common. In concrete structures whose surfaces were to remain unplastered or unclad, as exposed concrete, the main types of defect and fault found were: chipping above the reinforcement, with rust patches, pockets of gravel, ridges and other surface flaws, as well as damaged edges. Slight colour differences, fine reticular cracks, pores and light sanding were other minor faults.

## Points for consideration

– Flaws, and particularly pockets of gravel, occur mainly if the maximum aggregate size is equal to, or greater than, the depth of concrete cover over the reinforcement, so that parts of the aggregate collect between the reinforcement and the formwork and prevent the voids from being filled up with fine mortar. Moreover, if the maximum aggregate size is too large in relation to the smallest dimension of the structure, it will be impossible to achieve sufficient compaction of the exposed concrete. Defective laying, compaction and secondary compaction will produce the same flaws.

– Insufficient cover over the steel reinforcement, e.g. as a result of failure to use spacers, or because the reinforcement has been trodden down (see B 1.1.4) are also causes of pockets of unmixed gravel and chipping, which expose the reinforcement.

– A flawless exposed concrete surface also depends on the nature of the formwork boards used (tightly joined, uniform thickness, pre-treated with release agents, see B 1.1.4), on the way in which the concrete is laid, and particularly on the arrangement and construction of joints in the work (see B 1.1.5) and the treatment of the concrete after laying (see B 1.1.6).

– The use of different types of cement and aggregate containing dirt which may be washed out (e.g. lumps of clay), results in discoloration and patches of dirt on the concrete surface.

## Recommendations for the avoidance of defects

● The maximum aggregate size ($\leq$ 15 mm) should be suited to the necessary concrete cover over the reinforcement, or should, for reasons of workability, be determined in accordance with the smallest dimension of the slab or beams ($\leq$ 1/5). In order to obtain perfect exposed concrete surfaces, particular attention must be paid to ensuring that the reinforcement is sufficiently covered.

● In order to avoid flaws in exposed concrete surfaces, the formwork boards should be laid as tightly as possible, must be clean, and treated with suitable release agents. In addition, the concrete must be well compacted (possibly compacted a second time), and given the correct after-treatment to suit the type of concrete and weather conditions. Care should be taken to ensure that joints in the work are arranged and constructed correctly.

● In order to avoid discoloration and soiling, the concrete should contain only the same types of cement and uniform clean aggregates.

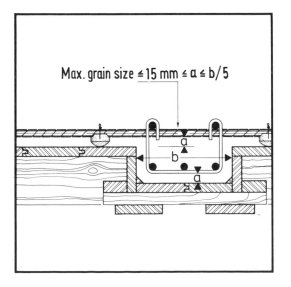

Max. grain size $\leq$ 15 mm $\leq$ a $\leq$ b/5

Ceilings
Surface coverings

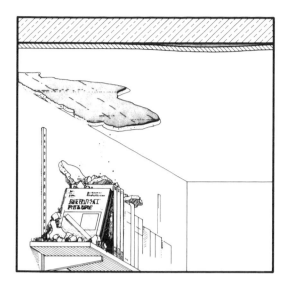

Instances where plaster comes away from reinforced concrete slabs represent a wide-ranging group of damage to internal surface coverings. In many cases, areas of plaster did not have good adhesion with the backing material – there were hollow areas. Some of the damage which occurred was severe, since large areas of falling plaster can represent a serious danger.

Lime, gauged lime, and gypsum plasters applied to reinforced concrete ceiling slabs and reinforced concrete floor slabs were affected by this type of damage.

**Points for consideration**

– Adhesion of the plaster to the substratum is of particular importance with finishes on the underside of reinforced concrete slabs, since the plaster is suspended from the substratum and the adhesive bond is placed under stress both by deformation of the reinforced concrete slab and of the plaster. Elastic, creep and shrinkage deformation, and – in the case of roof slabs and ceilings with surface heating – thermal expansion and contraction, produce stresses in the bond and, in the absence of an edge joint around the perimeter of the slab/ceiling, compressive stresses in the plaster layer.

– The physical and chemical processes that are operative in the adhesion of plaster have not been fully explained and there are major differences between the different plaster groups (e.g. gauged lime and gypsum plasters) and any agents added to promote adhesion. Nevertheless, a few general comments can be made on which properties of the substratum have an adverse effect on the adhesion of the plaster:

– Moisture content of the concrete too high:
Depending on drying conditions (weather, ventilation), it takes between 2 and 8 weeks from the time the formwork is struck, before the substratum is sufficiently absorbent. This absorbency is important, both in terms of the adhesion of the newly applied plaster (adhesion of the mixing water to the substrate) and for allowing the plaster to key with the substratum (suction of binding agent solution into the pore system of the plaster substratum). Plaster adhesion can also be reduced by the salts transported to the surface between the concrete and the plaster, because of the increased quantities of water given off by concrete which is plastered too early. Concrete should therefore be allowed to dry sufficiently before plastering is commenced.

– Loose particles on the surface of the concrete:
Loose parts of the concrete surface itself – e.g. concrete slurry, and layers of sintered material, especially in the case of prefabricated structures – as well as other foreign bodies, e.g. splashes of mortar, or dust, adhere only loosely to the underside of the slab and can therefore cause the plaster which is applied to spall from the slab, together with these particles. They must therefore be removed before plastering begins.

– Release agent residue on the concrete surface:
Sometimes there may be such a thick residual layer of release paste on some areas of the concrete surface (especially as a result of uneven application) that adhesion between the concrete and the plaster is severely reduced. Thick residual layers of a release agent – which can be recognised if water applied in a wetting

Ceilings
Surface coverings

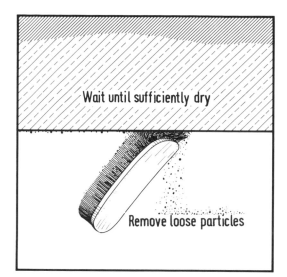

trial is repelled – must be removed by aqueous solvents. When using thin adhesion plasters, particular attention must be paid to obtaining a very smooth concrete surface, otherwise some areas of the plaster may be very thin. The development of the strength of these very thin plasters depends greatly on the absorbency of the substratum and on the drying conditions in the room; they therefore tend to crack.

– Because of the importance of the condition of the concrete surface for obtaining damage-free plasters, good building practice requires that the plastering contractor tests the plaster substratum, and if necessary, records in writing any reservations he may have.

### Recommendations for the avoidance of defects

● Before plastering reinforced concrete slabs, the condition of the plaster substratum must be carefully checked. In particular, the moisture content, absorbency, strength and even surface of the concrete must be checked by close inspection, and by cleaning, scratching and wetting tests.

● Where the substratum is still wet, wait until the concrete has dried further. Loose particles must be removed by brushing, or sand blasting. Release agent residue should be removed by means of aqueous solvents.

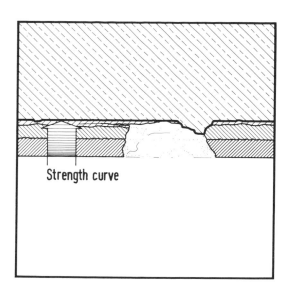

Strength curve

Even where the concrete surface was perfect, ceiling plaster was still found to spall from reinforced concrete slabs where – when using gauged lime or gypsum plasters – the surface had not been pre-treated with a daubing coat, or if the low strength daubing coat had been applied so smooth that it did not provide sufficient adhesion to the plaster coat.

In thin single coat plasters (adhesion plasters), the surface coverings were found to spall from the substratum where there were wide fluctuations in the thickness of the plaster, and where the underside of the ceiling was not smooth. Major strength losses and spalling were also found in gypsum plasters in frequently used wet rooms (shower rooms).

**Points for consideration**

- The daubing coat used under reinforced concrete slabs serves to enlarge the adhesion surface and improve the key between the substratum and the plaster. The daubing coat can achieve this only under the following conditions:

    - The strength of the daubing coat must not be any less than that of the plaster coat to be applied to it – the proportion of binding agent in the daubing mortar must therefore not be any less than that of the plaster mortar, and the binding agent used must be suitable for the type of mortar.

    - The daubing coat must produce a rough surface – it should therefore contain a high proportion of coarse sand (diam. 5 mm).

    - The daubing coat must have sufficient strength before plastering begins    at least 12 hours should therefore elapse between application of the daubing coat and plastering.

- When added to the mortar for the daubing coat, or if used instead of the daubing coat, fine plastic aggregate (e.g. on a PAC base) can, as the surveys have shown, increase the adhesion of the mortar particles between one another, and to the substratum (because the plastic molecules form a film), and also increase their resistance to deformation without cracking because of the elasticity of the plastic film. However, this type of admixture is not standardised. The same is also true of thin, single-coat 'adhesion plasters' – gypsum plasters from mortar group IV, made from ready mixed dry mortars, with adhesion promoters, and generally with a cellulose base.

    Single coat adhesion plasters do not generally have the 15 mm thickness required under Codes of Practice. Their use is therefore subject to special agreement. Their slenderness also necessitates increased and specially agreed requirements in terms of the uniformity of the substratum, as some areas of the plaster coat may be only a few millimetres thick and will tend to spall. Single coat adhesion plasters should therefore be no thinner than 8–10 mm.

- Gypsum mortar hardens because needle-shaped crystals form which knit with one another and with the admixtures. These crystals are soluble in water. Thus, if a gypsum mortar remains soaked for lengthy periods, it will lose its strength. Gypsum plaster is therefore not suitable for rooms of high humidity content.

Ceilings
Surface coverings

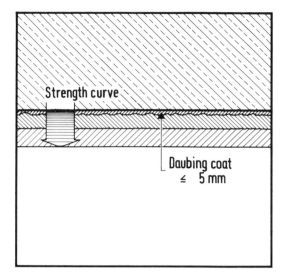

Strength curve

Daubing coat
≤ 5 mm

### Recommendations for the avoidance of defects

● In principle, ceiling plaster should be applied to a substratum which has already been treated with a daubing coat. The strength of the daubing coat should be at least equal to that of the coat of plaster to be applied to it. The aggregate used should be sand with a high proportion of coarse grains (diam. 5 mm), and it should be applied at least 12 hours before plastering begins.

● Arrangements for plaster types and plastering methods – such as the use of adhesion promoting agents, or single coat adhesion plasters – should be agreed separately. Thin, single coat adhesion plasters should be used on ceilings with fine surface tolerances and should not be less than 8 mm thick.

● Gypsum plasters should not be used in rooms exposed to large amounts of moisture.

Ceilings
Surface coverings

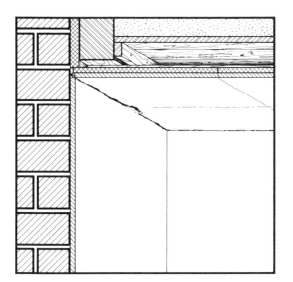

Particularly serious and dangerous failures occurred if plaster supporting materials were inadequately fixed to the ceiling structure – e.g. with nails that were too short – and if they parted from the ceiling structure over large areas.

Cracks were also common at the joints between plaster supporting panels – generally woodwool – and multi-layer lightweight building slabs. In the case of reinforced concrete ceiling slabs, where the panels were fitted in the area of the window lintels and the ceiling supports to prevent a thermal bridge, cracks formed along the edge between the lightweight building panels (serving as a backing for the plaster) and the reinforced concrete.

In examples of timber joist ceilings – generally pitched roof ceilings – the edges of the lightweight building panels, used as supporting material for the plaster, were visible as cracks. In all these examples, there was nothing to strengthen the joints between the panels.

**Points for consideration**

– Especially if plaster supporting materials are fitted to timber (battens, joists) with nails, defective attachment becomes apparent not merely when the plaster is applied, but can also cause the whole of the supporting structure to part dangerously if, 1 to 2 years after the building is first occupied, the nails begin to lose their grip as the timber shrinks.

– In the case of gypsum materials, lightweight woodwool building panels with an unacceptably high proportion of water-soluble chloride (when used on structures which are liable to become soaked by condensation, e.g. ceilings above wet rooms), with attachments which have been given inadequate corrosion protection, may result in ceiling plaster spalling from the supporting material several years after it has been applied.

– Because of the importance of the defects which occurred, specific installation instructions and guidelines are recommended when using a plaster finish to supporting panels made of lightweight foamed plastic, woodwool and suspended ceilings of plasterboard. Thus, for example, lightweight woodwool building panels 25 mm thick and 500 mm wide must be attached to a timber supporting structure with at least three galvanised lightweight building panel nails (3.1 × 60 mm) at intervals of no more than 670 mm.

– Lightweight woodwool building panels are susceptible to a large amount of shrinkage and expansion (1.5 – 3.5 mm per m). In the case of panels which have been stored damp and then fitted in a swollen state (e.g. packing lost), the contraction which takes place when the panels dry out results in tension and cracking in the plaster. Although wire mesh strips embedded in the daubing coat along the joints and edges of the panels are not able to prevent them moving, they nevertheless distribute the stresses over a wider area of the plaster surface and thus prevent the visible cracking in internal plasters, which are subject to lower stresses than external plasters.

Ceilings
Surface coverings

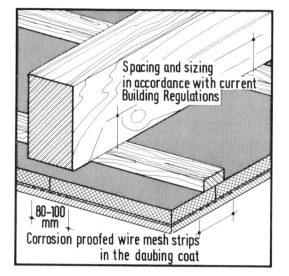

Spacing and sizing
in accordance with current
Building Regulations

80-100
mm

Corrosion proofed wire mesh strips
in the daubing coat

## Recommendations for the avoidance of defects

● When securing plaster supporting materials to the underside of the ceiling, the requirements of the relevant standards in terms of type, minimum dimensions, method of installation and corrosion resistance of the means of attachment must be taken into account. Careful checks should be made during construction to ensure that these requirements are being met.

● When lightweight building panels are laid together and plastered dry, all the edges of the panels must be provided with corrosion-proofed wire mesh strips at least 80–100 mm wide, which are embedded in the daubing coat.

Ceilings
Surface coverings

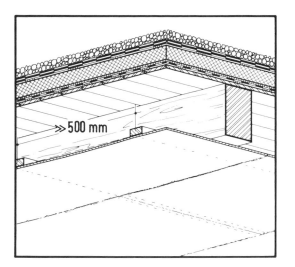

In gypsum plasterboard panels fitted as a ceiling cladding to timber joist ceilings, cracking was observed at the joints between the panels and around the edges of the ceiling.

Cracking and severe deflection occurred if the supporting structure for the panels was too widely spaced (e.g. if the panels were fitted direct to the timber joist).

**Points for consideration**

– Structures which support ceilings are subject to deformation which is dependent on loading and moisture. This deformation and movement is transmitted to the gypsum plasterboard panels and results in cracking in the surface coverings above the joints between the panels if these are not covered with 'reinforcing strips'.

– Covering strips, which can be inserted between panels with bevelled edges, serve to provide a continuous transition between the edges of the panels and thus to distribute the differences in the movements of the panels over a larger area. This method, however, only permits bridging of cracks up to a maximum of about 0.5 mm and is not suitable for the large differential movement which can, for example, occur along the edges between timber joist ceilings and load-bearing brickwork.

– In order to ensure secure attachment and to prevent harmful deflection of the plasterboard panels, current Building Codes give precise details of the type, size, spacing and attachment of the supporting structure and of the plasterboard panels. Thus, for example, the maximum permissible span for 12.5 mm thick plasterboard panels is 500 mm in the case of transverse attachment. When using roof battens (24 × 48) as the supporting structure, panels should be attached at intervals of no more than 650 mm (e.g. by fitting backing battens) in order to keep deflection below span/1000.

**Recommendations for the avoidance of defects**

● In examples of ceiling claddings, made of plasterboard, which are to be painted or papered, panels with bevelled, cardboard-covered edges should be used. All these edges should be filled after inserting 'reinforcing strips'. Joints should be arranged around the perimeter of the ceiling (e.g. separation using an adhesive strip which is not covered with filler) so that the anticipated linear expansion which will occur betweeen the ceiling and the wall will not lead to uncontrolled cracking.

● When selecting the type and size of the attachments to be used, and the type and size of supporting structure, the provisions of the current Codes of Practice should be taken into account.

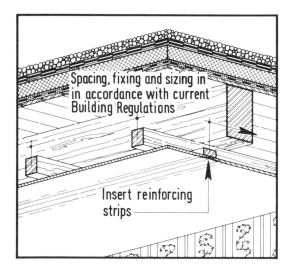

Ceilings
Surface coverings

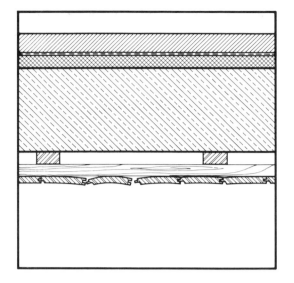

Where tongued and grooved boards were used as the cladding for timber joist ceilings and reinforced concrete ceiling slabs, wide shrinkage cracks and warping occurred and there were open joints between the boards. In many cases, the boards were fixed in such a way that their movement was restricted. In some of the defective cladding, the shrinkage contraction due to the fact that the boards were fitted whilst damp was so great that, even though the boards were fixed correctly, they freed themselves from the grooves and caused open joints and deflection.

## Points for consideration

– By absorbing and giving off moisture, cellular wood fibre expands and contracts and results in the swelling and shrinkage of wood in a range between full saturation (30 wt. %) and complete dryness. This deformation occurs almost exclusively radially and tangentially to the line of the trunk. Tangential deformation and radial deformation, for example, are 30 and 20 times greater, respectively, than that along the axis of the trunk. These different deformation characteristics, as well as differences between heartwood and sapwood, result in uneven deformation, especially in the case of boards.

– Apart from the once-and-for-all drying which occurs when timber is first cut, the moisture content of wood fluctuates constantly in accordance with the moisture content of the surrounding air and thus, for example in a normal living room, varies between 7 and 17% wt. (average balanced moisture content is 12% wt.).

– Since timber for carpentry work with a moisture content of 20% wt. is regarded as 'dry' and is therefore suitable for internal claddings, additional shrinkage can generally be expected in newly fitted claddings until balanced moisture content is attained. However, if the timber is stored for a reasonably long period under the same conditions as those in which it is to be installed, its moisture content will approach the average balanced moisture content actually before it is fitted. In this way, the shrinkage deformation of the fitted boards can be reduced.

– Immediately upon completion of a building, reinforced concrete ceiling slabs generally still have a high moisture content. This moisture is given off to the surrounding air and adjacent building materials and can thus result in swelling of timber boards fitted directly beneath them. In this situation, therefore, the shrinkage deformation of the fitted boards will be particularly high until average balanced moisture content is attained.

– If the boards are fitted in such a way as to prevent, totally, any moving, in the event of shrinkage or swelling deformation (e.g. by nailing them on all sides) shrinkage cracks and/or warping will occur even in adequately dry timber once the means of attachment have been removed. Thus, the timber should be attached so that its swelling and shrinkage movements are restricted only sufficiently to prevent deformation. These requirements can, for example, be met by tongued and grooved boards secured to the supporting structure on one side by means of brackets in the groove.

Use seasoned timber
Install as late as possible

### Recommendations for the avoidance of defects

– The moisture content of timber boards for ceiling claddings should, wherever possible, be equivalent to 'balanced moisture' before they are installed. They must, therefore, at least be 'dry'; moreover, it is also advisable to allow them to adapt to the anticipated room conditions by storing them in the room for a reasonable period of time.

– In new buildings, timber layers should be fitted beneath reinforced concrete ceilings as late as possible after the ceiling has been completed.

– The way in which the boards are secured must prevent them from warping, but should also impede movement of the timber as little as possible. Indirect fitting by means of special brackets is a particularly good solution.

Ceilings
Surface coverings

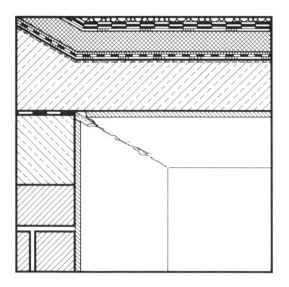

Cracking in surface finishes was most frequently found around the edges of the ceiling. In all cases, there was no special plaster or filler in the joint between plaster or plasterboard surfaces and the wall surface. Cracking was particularly common in reinforced concrete ceiling slabs and timber joist ceilings.

## Points for consideration

– Because of the loading resulting from dead load and imposed load, and as a result of shrinkage, reinforced concrete ceiling slabs are subject to elastic and plastic deformation which can lead to pressure at the inner edges of the supporting structure. Moreover, where the length of restraint is short, or where there is no edge reinforcement, reinforced concrete ceiling slabs can work loose in the corners of the building as a result of lifting of the supports. Similarly, in examples of reinforced concrete ceiling slabs – with displaceable supports to absorb shrinkage and thermal expansion – movements of the ceiling, in relation to the wall, are to be expected.

– Deformation is also to be expected in timber joist ceilings as a result of deflection caused by loading, and shrinkage and swelling caused by fluctuations in moisture content.

– If the transitions between the ceiling surfaces and the wall surfaces are simply plastered together or filled, the above mentioned deformations will result in tensile and compressive stresses along the edge of the connection, and cracking and spalling of the surface finishes. If, however, the ceiling plaster is separated from the adjacent structures by means of an edge joint, it is then able to follow the movements in the ceiling supporting structure unhindered. The size and construction of the joint depend on the extent of the deformation anticipated.

## Recommendations for the avoidance of defects

● In principle, the ceiling plaster should be separated from adjacent wall surfaces by means of a joint. Where the anticipated movement is slight, it is sufficient if, before plastering, the ceiling and wall plaster are separated from one another, e.g. by means of a trowel cut or by using galvanised plaster stop. In the event of considerable movements (e.g. roof ceilings on flexible mountings) the width of the joint must be matched to the expected linear deformation. Subsequent filling with solid material should be prevented by installing two-piece plaster cove or covering beads.

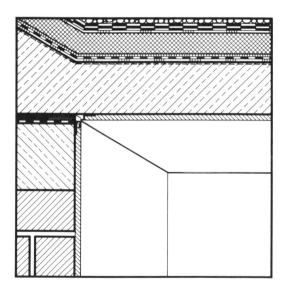

### Exposed concrete

– Specialist books and guidelines

Rapp, G.: Technik des Sichtbetons. Beton-Verlag, Düsseldorf 1969.
Wesche, K.: Baustoffe für tragende Bauteile, Band 2, Beton und Mauerwerk. Bauverlag, Wiesbaden und Berlin 1974.
DIN 18331 – VOB, Teil C.: Beton- und Stahlbetonarbeiten. September 1976.

– Specialist papers

Grunau, E. B.; Menzel, H.: Sanierung von Sichtbetonflächen. Sonderdruck aus: Kunststoffe im Bau, Themenheft 26.
Grunau, E. B.: Strukturbeton durch Kunststoffmatrizen. In: Das Baugewerbe, Heft 2/1973, Seite 24–33.
Köneke, R.: Betonschutz aktuell. In: Das Baugewerbe, Heft 1/73, Seite 22–25, Heft 2/73, Seite 22–23; Heft 3/73, Seite 42–44; Heft 5/73, Seite 91–98; Heft 6/73, Seite 41–42; Heft 7/73, Seite 40–41; Heft 9/73, Seite 69–70; Heft 15/73, Seite 32–34.
Schmidt-Morsbach, J.: Schalung und Betonoberflächen für Betonfertigteile. In: Betonwerk + Fertigteil-Technik, Heft 3/1974, Seite 177–190.
Schmincke, P.: Frischbeton-Konsistenz und Sichtbeton. In: Das Baugewerbe, Heft 2/1976, Seite 22.
Schwara, H.: Beton-Sichtflächen. In: Das Baugewerbe, Heft 7/1973, Seite 33–38.
Zement-Merkblatt »Sichtbeton«. Herausgeber Bundesverband der Deutschen Zementindustrie e.V.

### Ceiling plaster

– Specialist books and guidelines

Albrecht, W.: Zusatzmittel, Anstrichstoffe, Hilfsstoffe für Beton und Mörtel. Bauverlag, Wiesbaden und Berlin 1968.
Piepenburg, W.: Mörtel, Mauerwerk, Putz, 6. Auflage. Bauverlag, Wiesbaden und Berlin 1970.
DIN 18350 – VOB, Teil C: Putz- und Stuckarbeiten, August 1974.
DIN 1101: Holzwolle-Leichtbauplatten; Maße, Anforderungen, Prüfung. April 1970.
DIN 1102: Holzwolle-Leichtbauplatten nach DIN 1101; Richtlinien für die Verarbeitung.
DIN 1104 Blatt 2: Mehrschicht-Leichtbauplatten aus Schaumkunststoffen und Holzwolle; Richtlinien für die Verarbeitung. April 1970.
DIN 4121: Hängende Drahtputzdecken; Putzdecken mit Metallputzträgern, Rabitzdecken; Anforderungen für die Ausführung, September 1968.
DIN 18550: Putz, Baustoffe und Ausführung. Juli 1967.
DIN 18550, Beiblatt: Putz; Baustoffe und Ausführung, Erläuterungen. Juni 1967.
DIN 18202 Blatt 2, Vornorm: Maßtoleranzen im Hochbau, Ebenheitstoleranzen für Oberflächen von Wänden, Deckenunterseiten und Bauteilen. Juni 1974.

Merkblatt des Bundesverbandes der Gips- und Gipsbauplattenindustrie e.V. und des Deutschen Stuckgewerbebundes im Zentralverband des Deutschen Baugewerbes: Gipsputz auf Beton.

– Specialist papers

Albrecht, W., Wisotzky, Th.: Über die Putzhaftung an Betondecken bei Verwendung von Haftvermittlern und Haftputzmörteln. FBW Blätter, Folge 3, 1968.
Albrecht, W., Wisotzky, Th.: Über die Härte von Innenputzen. In: bau + bauindustrie, 10/1968.
Grunau, E. B.: Neue Kunstharzdispersionen als Zusatz für Mörtel und Putze. In: Baugewerbe 5/78.
Hauck, W.: Haftputz auf glatt geschalter Stahlbetondecke, Ablösung durch unsachgemäßes Aufbringen oder durch Einbau von Gußasphaltestrich? In: DAB 6/74.
Könecke, R.: Putzhaftungsschwierigkeiten an Betondecken. In: Das Baugewerbe 5/1975.
Oswald, R.: Schäden an Oberflächenschichten von Innenbauteilen. In: Forum-Fortbildung Bau, Heft 9/1978. Forum Verlag, Stuttgart.
Poch: Haftung von Gipsputz auf Beton. In: Innenausbau, 1/1976.
Schumann, D.: Untersuchung der bauphysikalischen Eigenschaften von Gips- und Gipsbauplatten. In: Bauwirtschaft, Heft 4/1972.
Volkart, K.: Neuzeitliches Putzen mit Gips. In: Das Baugewerbe, 6/1970.

### Timber claddings

– Specialist books and guidelines

Wesche, K.: Baustoffe für tragende Bauteile 4, Holz, Kunststoffe. Bauverlag, Wiesbaden und Berlin 1973.
DIN 18334 – VOB, Teil C: Zimmerarbeiten. Dezember 1958.
DIN 68365: Bauholz für Zimmerarbeiten, Gütebedingungen. November 1957.
Arbeitsgemeinschaft Holz e.V.: Merkblatt Hinweise für das Anbringen von Profilbrettern. In: Informationsdienst Holz.

### Gypsum plasterboard panels

– Specialist books and guidelines

Hanusch, H.: Gipskartonplatten, Trockenbau, Montagebau, Ausbau. Verlagsgesellschaft Rudolf Müller, Köln-Braunsfeld 1978.
Scheidemantel, H.: Handbuch der Plattentechnik; Bauplanung – Bauausführung – Bauüberwachung. Verlagsgesellschaft Rudolf Müller, Köln-Braunfeld 1969.
DIN 18180: Gipskartonplatten, Arten, Anforderungen, Prüfung. Juni 1967.
DIN 18181: Gipskartonplatten im Hochbau, Richtlinien für die Verarbeitung. Januar 1969.

– Specialist paper

Jäger, R.: Gipskartonplatten unter Brettbindern. Verwölbungen und Rißbildungen infolge fehlerhafter Unterkonstruktion. In: Bauschäden Sammlung, Band 3, Seite 118–121, Forum-Verlag, Stuttgart 1978.

## Problem: Floating cement screeds

Floating screeds are the most common type of floor construction in buildings with residential accommodation, since such rooms make the highest demands on the ceiling and floor structures in terms of their structural protection.

Storey slabs generally have to provide protection against air-borne and impact sound if they separate living or working rooms. In addition, floors have to provide thermal insulation when they separate heated rooms from the open air (opening), from the ground (basement floor), or from adjacent rooms which are not heated. In addition, underfloor heating means that it is necessary to ensure that not too much heat is lost to the room below.

With a floating screed construction laid on sound damping materials it is possible to improve sufficiently the protection offered by storey slabs against air-borne noise and impact noise. On all sides, the impact noise sound damping layer separates the screed from adjacent structures. This produces a double layer floor structure. Sufficient thermal insulation can also be provided at the same time by including thermal insulation layers of appropriate thickness.

Unlike bonded screeds, floating screeds, which are laid on flexible soft insulating materials, have a load distributing effect and are exposed to deflection. The thickness and strength of the screed must therefore be matched to the expected loads and the compression characteristics of the insulating layer. Moreover, additional demands are made on floating screeds – as on floor linings and floorcoverings – in terms of freedom from cracks, surface hardness and being level.

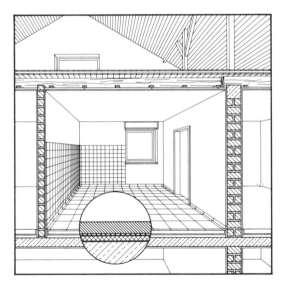

The largest number of defective floors were of the floating screed type because of the latter's widespread use. This section deals exclusively with this type of floor, although many of the defects discussed can be applied to other screed materials.

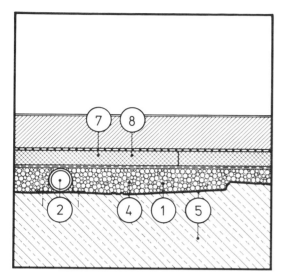

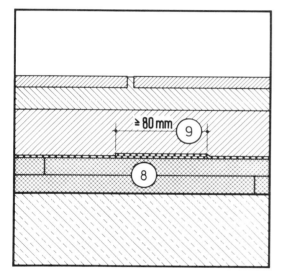

**1** The substratum for floating screeds must be suitably level. Spot irregularities must be removed and large irregularities should be covered with a filler course (see C 1.1.2).

**2** Service pipes, such as conduits for electrical cables or service pipes, must be secured to the surface of the slab and embedded in the filler course in such a way as to produce a firm and level surface (see C 1.1.2).

**3** If, however, it is necessary to lay service pipes, etc., inside the layers of impact sound insulation material, these should be arranged around the perimeter of the screed slab and should be covered with high-efficiency insulating mats or felts. The design depth of the pipes, etc., including the covering of insulating material, should not exceed the depth of the layer of insulating material (see C 1.1.2).

**4** The filler course can take the form either of a smoothing render or of a loose fill of light – not highly hygroscopic – materials which can be compacted and then covered. Loose sand fills should not be used (see C 1.1.2).

**5** Before a floating screed and a surface covering are laid, the whole substratum (generally the ceiling slab or a filler course) must be sufficiently dry throughout its whole cross-section and not merely on the surface (see C 1.1.2).

**6** The sound insulation provided by ceiling structures below and above living rooms must be determined according to the use made of the 'noisy room' and the maximum levels of the 'quiet rooms'. Thus, for example, in living rooms which have recently been completed, the air-borne sound insulation level should be at least 3 dB (5dB is better) and the impact sound insulation level at least 13 dB (20 dB is better) (see C 1.1.3).

**7** Only insulating materials whose suitability has been proven can be used as insulating layers for floating screeds (see C 1.1.3).

**8** The insulating layer should be laid without voids over the whole of the surface, pressing the offset seams together. It is preferable to lay the materials in two layers rather than one (see C 1.1.3).

**9** The insulating layer should be completely covered with strips of sufficiently strong water-resistant material to form a level surface. The seams of the covering should be overlapped by at least 80 mm. Suitable materials are, for example, 333 grade non-sanded bituminous felt or polyethylene films, 2 mm thick (see C 1.1.3).

Floors

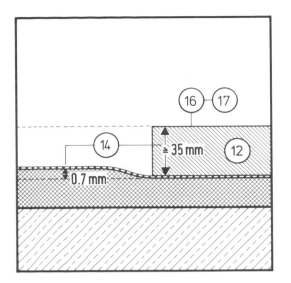

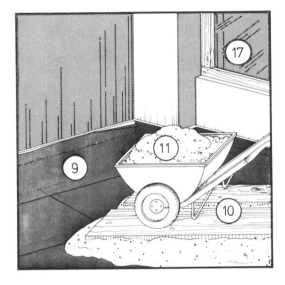

**10** During transportation and laying of the screed the insulating layer and the covering layer should not become displaced or damaged. They should be protected with boards to distribute the load (see C 1.1.5).

**11** Care should be taken to ensure that the screed mortar is correctly and uniformly made up and that it is homogeneously mixed in a concrete mixer for at least three minutes. It should consist of a washed aggregate with a maximum grain size of 8 mm, with between 350 and a maximum of 400 kg of cement per m$^3$. The water/cement ratio should not exceed 0.6 and the consistency should be between stiff and plastic (see C 1.1.4 and C 1.1.5).

**12** The average strength of the screed should not fall below the values specified in the Codes of Practice (see C 1.1.4).

**13** The maximum continuous area of the screed should not exceed 30 m$^2$, or 20 m$^2$ in the case of screeds over underfloor heating systems; similarly, the maximum length of an area of screed should not exceed 6 m, or 5 m in the case of screeds over underfloor heating systems, or screeds which are exposed to strong sunlight (see C 1.1.4 and C 1.1.8).

**14** The minimum thickness of the screed depends on the compression characteristics of the insulating layer. In the case of the typical loads found in residential buildings, and taking into account the compression characteristics of the insulating layer, it should be between 35 and 45 mm thick. (CP 204 Part 2 recommends in UK a thickness not less than 65 mm).

**15** Floating screeds should be compacted as carefully as possible, and preferably with a surface vibrator (see C 1.1.5).

**16** The compacted surface of the screed should be trowelled level over accurately adjusted guides and should then be smoothed. The tolerances specified should be observed according to the particular application concerned (see C 1.1.5).

**17** The finished screed must be kept damp for a period of at least 7 days and should then be kept from drying out for a further period of at least 14 days (see C. 1.1.5).

**18** Screed additives, such as synthetic resin admixes, can be used to improve workability, or increase the water retention capacity of the mortar, thus simplifying the necessary protection measures (see C 1.1.4 and C 1.1.5).

**19** Effective measures must be taken to prevent access to freshly laid screed surfaces (see C 1.1.5).

**20** Floor surfaces that are in contact with the soil, and which are to be covered with a floating screed, must have a damp-proof and capillary-tight seal beneath the layer of insulating material. This can be provided by two layers of sealing membrane bonded over the whole of their surfaces, or single layer heavy duty plastic film with welded seams; these should then be tightly joined to the horizontal wall damp-proof course (see C 1.1.6).

**21** The thermal resistivity of floors in heated rooms which are in contact with the ground should be at least 0.86 m$^2$ K/w (k $\leq$ 0.97 W/m$^2$K) (see C 1.1.6).

**22** In rooms with increased air humidity a vapour barrier should be fitted above the insulating layers, with a low level of vapour resistance (see C 1.1.6).

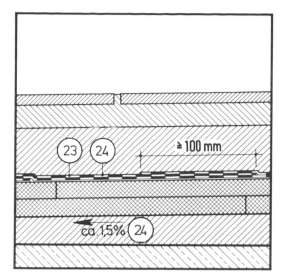

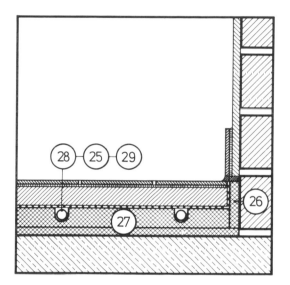

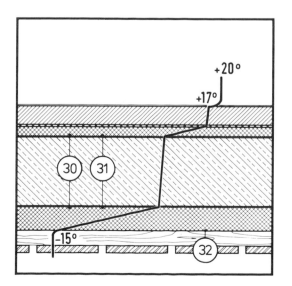

**23** In damp rooms in residential buildings which are not heavily exposed to moisture, a single layer damp-proof course with bonded seams can be laid directly on top of the insulating layer (see C 1.1.7).

**24** However, if intense water exposure is expected, the damp-proof course must consist, for example, of two layers of sealing membrane bonded together over the whole of their surfaces and laid with a fall of 1.5%; they should also, like the floor surface, be linked to a floor drain. The fall should be obtained either by a sloping screed or by a compacted filling (see C 1.1.7).

**25** The heating elements for underfloor heating systems should be laid in their own bed which forms a level and suitably strong basis for the screed. Suitable beds are, for example, layers of thermal insulation materials with channels of sufficient depth as well as screed layers (see C 1.1.8).

**26** Expansion joints of suitable dimension should be placed along all perimeters and supports and surround all projections. Floors with surface areas in excess of 20 m² should be subdivided by means of expansion joints. The individual floor areas produced in this way should then be provided with separate heating circuits (see C 1.1.8).

**27** The thermal insulation layer beneath underfloor heating systems should have a thermal resistivity of $\geq 1.25$ m² K/W (see C 1.1.8).

**28** Underfloor heating systems should not be put into operation until the screed and mortar or adhesive layers of surface finishes have fully hardened and dried out. In the case of cement screeds, at least 28 days should be allowed after laying. After this, the temperature should be raised gradually, in daily intervals of about 10 K (see C 1.1.8).

**29** The pipes used for underfloor heating systems should be manufactured of corrosion resistant material or should be provided with lagging to protect them from corrosion (see C 1.1.8).

**30** The thermal resistivity of slabs above unheated basements must be at least 0.86 m² K/W (thermal conductivity coefficient $\leq 0.97$ W/m² K) (see C 1.1.9).

**31** The thermal resistivity of slabs which are in contact with the open air must be at least 1.72 m² K/W (thermal conductivity coefficient $\leq 0.49$ W/m² K) (see C 1.1.9).

**32** Where possible, the necessary layer of thermal insulation should be fitted to the underside of the ceiling slab. If, however, it is necessary for it to be fitted to the upper side, the thermal insulation materials used should have a high resistance to vapour. Thermal insulation materials which have a low vapour resistance (e.g. glass wool) must have a vapour barrier on their upper surface in the form of laminated aluminium foil or similar (see C 1.1.9).

Floors
Floating cement screeds

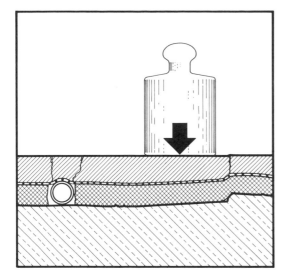

Various types of defects in floating screeds were attributable to defects in the substratum.

Cracks and breaks occurred in screeds above relatively large irregularities in the base, which caused weaknesses in the screed. Cracks and damage to screeds were also detected where day joints or underfloor heating pipes were laid on the unfinished slab and in the compaction of insulation layers in such a way that the thickness of the screed was reduced. In some of these cases, the level of impact resistance was at the same time lower than required. Signs of damp appeared and insulation layers made of vegetable fibres rotted if the floating screed, and subsequently the floor surface covering, was laid too early on very damp reinforced concrete slabs.

**Points for consideration**

– In floating screeds, the condition of the substratum (which generally takes the form of a reinforced concrete slab) is of great importance, both in terms of the impact resistance to be obtained and in terms of ensuring that the screed and the surface coverings remain free of defects.

– However, although the whole surface of the substratum is covered with a layer of insulating material, because the latter is so thin and elastic, even isolated irregularities in the surface of the substratum will have the effect of reducing its thickness and thus of reducing impact noise insulation and resistance.

– Irregularities in the base slab, e.g. projections between different sections of concrete, which amount to more than 8 mm in 1 m, result in considerable fluctuations in the thickness of a screed which is only between 35 and 45 mm thick. Where the screed is also subject to linear deformation as a result of shrinkage or spot loads, this can also result in cracking.

– Deviations in the horizontal measurements, e.g. as a result of ceiling deflections or projections, or deviations in the dimensions of the unfinished base slab cannot be evened out by means of screeds. Instead, it is necessary to use special smoothing renders.

– Services which are laid on the base slab represent major irregularities. However, they pose no problem provided they are embedded in a filler course in such a way as to form a solid and level surface. Any adverse effect on the insulating layer can also be avoided if the depth of the service pipes or wires, including a cover made of a specially high-grade, elastic insulating mat ($s \leq 3 \times 10^7$ N/m³ (3 kg/cm³)), is equivalent to the depth of the insulating layer, so that a reduction in screed thickness and acoustic bridging are avoided.

– Newly cast reinforced concrete slabs, as well as wet laid levelling courses retain a considerable amount of water for relatively long periods throughout their depth, even though they appear dry on the surface. Indeed a fairly high water content is to be expected in buildings which have been complete for some time if they repeatedly become soaked with precipitation.

– If a floating screed is laid on a supporting structure which is still damp, diffusion of the water will be greatly retarded and – especially where there are temperature differences between the individual layers – the moisture may move to between the floor substratum and the insulating or the screed.

Floors
Floating cement screeds

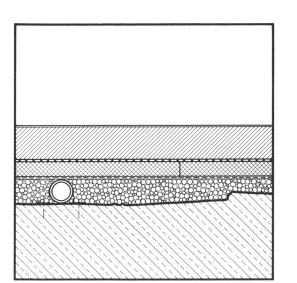

Under the influence of increased moisture levels, non-moisture-resistant insulating layers will rot, adhesives will swell and wooden parquet flooring will warp.

### Recommendations for the avoidance of defects

● The substratum for floating screeds must be suitably level. Spot irregularities must be removed and large areas of irregularities must be covered over with a filler course. In the case of thin and particularly soft insulating layers, it may be necessary to impose further limits on the tolerances.

● Service pipes, such as conduits for electrical cables, must be secured to the surface of the slab and embedded in the filler course in such a way as to produce a firm and level surface.

● If, however, it is necessary to lay service pipes, etc., inside the layer of impact sound insulation material, these should be arranged around the perimeter of the screed slab and should be covered with high efficiency insulating mats or felts ($s \leq 3 \times 10^7 N/m^3$ (3 kg/cm³)). The design depth of the pipes etc., including the covering of insulating material, should not exceed the depth of the layer of insulating material.

● The filler course can take the form either of a smoothing render or of a loose fill of light materials (not highly hygroscopic) which can be compacted and then covered. Loose sand fills should not be used.

● Before a floating screed and a surface covering are laid, the whole substratum (generally the ceiling slab or a filler course) must be sufficiently dry throughout its whole cross-section, and not merely on the surface. The period allowed for drying out must be selected in accordance with the moisture content and thickness of the structures, and with the drying conditions (temperature, relative air humidity). Further exposure to soaking, for example as a result of precipitation, must be avoided.

Floors
Floating cement screeds

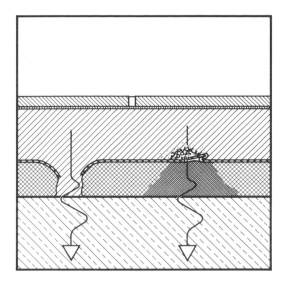

In some cases, floating screeds on reinforced concrete slabs did not attain the planned degree of insulation against impact sound. This was the case if the dynamic rigidity of the impact sound-insulating layer was too great because it was made of thermal insulation materials. Panels, matting or felt, which were not laid tightly and without voids, also had an unfavourable effect. At individual points, the screed mortar was in contact with the base slab, or the screed mortar or binding agent had penetrated open joints in the covering of insulating material so that the insulating layer had hardened, or mortar bridges had formed between it and the reinforced concrete slab.

**Points for consideration**

– The standard of insulation of ceiling structures against air-borne and impact sound must be determined in individual cases according to the type of use to be made of the room – the type and intensity of impact sound generation. Code of Practice 3: Chapter III: Part 2: 'Sound Insulation and Noise Reduction' sets out minimum values for the insulation of dwellings against air-borne and impact sound. With an impact sound insulation level of 3 (0) dB (minimum requirement) people will clearly be heard walking on the floor above; with an impact sound insulation level of 13 (10) dB (increased sound insulation) people could barely be heard walking on the floor above; whilst with a level of insulation in excess of 20 dB, normal sounds of this type were no longer audible. With an air-borne sound insulation level of $\leq 3$ dB, living noises (radio, speech) are damped sufficiently to ensure that they do not cause disturbance at night.

– Both the air-borne sound insulation and impact sound insulation of base slabs can be improved with a floating floor construction.

– The insulating layer of a floating screed separates the screed from the substratum, thus producing a double skin structure in acoustic terms. As the thickness of the screed increases and as the dynamic rigidity of the insulating layer decreases, the insulation provided by the screed against impact and air-borne noise increases.

– Dynamic rigidity 's' signifies the spring characteristics of the intermediate layer when a dynamic (alternating) load is imposed. It is directly proportional to the modulus of elasticity of the material and inversely proportional to the thickness of the layer. However, it is not possible to increase the thickness of the insulating layer or screed at will, since this in turn increases the depth of the structure and the loads involved.

– Dynamic rigidity values of less than $9 \times 10^7 \text{N/m}^3 (9\,\text{kg/cm}^3)$ in base slabs whose air-borne sound insulation is already adequate (less than $3 \times 10^7$ N/m³) in unfinished ceilings whose air-borne and impact noise insulation has to be improved, can only be achieved using impact sound insulating materials. Thermal insulating materials made of hard foamed plastic, which because they are profiled, do not significantly improve impact sound insulation, are considerably more rigid, so that it is normally necessary to fit an additional impact noise insulating layer. Checks should be carried out of each individual case (see C 1.1.9: thermal insulation of floors).

Floors
Floating cement screeds

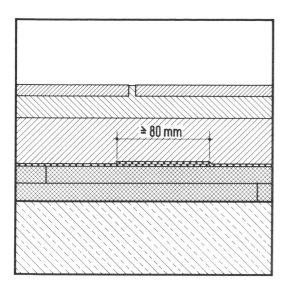

– When the screed mortar is laid, holes and open joints in the insulating matting or panels will cause the matting to be pressed down to the base slab by the weight of the mortar. Moreover, the screed mortar, or binding agent, can also penetrate through any seams in the insulating layer that have insufficient overlap, thus soaking the insulating layer and considerably increasing its dynamic rigidity. In either case, even individual flaws will result in a clear deterioration of the insulating effect of the floating screed structure.

– Folds in the covering material – especially when the latter takes the form of relatively stiff material such as 500 grade felt – will reduce the thickness of the screed and can also result in cracking. The risk of this happening can, however, be reduced by laying the insulating material in two layers with overlapping joints and, if the covering layer consists of tear-resistant material without flaws, with joints that are overlapped by at least 80 mm.

### Recommendations for the avoidance of defects

● The sound insulation provided by ceiling structures below and above living rooms must be determined according to the use made of the 'noisy' room and the maximum levels of the 'quiet' room. Thus, for example, in living rooms which have recently been completed, the air-borne sound insulation level should be at least 3 dB (5 dB is better) and the impact sound insulation level at least 13 dB (20 dB is better).

● Only insulating materials whose suitability has been proved can be used as insulating layers for floating screeds.

● The insulating layer should be laid without voids over the whole of the surface, pressing the offset seams together. It is preferable to lay the materials in two layers rather than one.

● The insulating layer should be completely covered with strips of sufficiently strong water-resistant material to form a level surface. The seams of the covering should be overlapped by at least 80 mm. Suitable materials are, for example, 333 grade non-sanded bituminous felt or polyethylene films, 2 mm thick.

Floors
Floating cement screeds

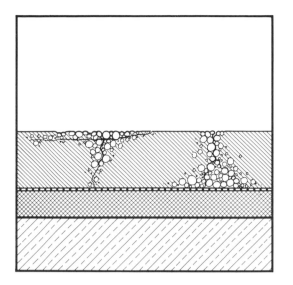

Fractures within the screed, loosening of tiles or parquet flooring, as well as complete decomposition at heavily loaded points, and cracks, could be attributed to inadequate strength of the cement screed. The screeds were highly porous, used aggregates which were too small, or contained insufficient quantities of cement.

Open cracks and fractured corners which went through the depth of the screed were observed if the finished floor covering was in general too thin, or if there were major fluctuations in the screed thickness, usually as a result of pipes laid on the base slab, large irregularities in the slab, insulating layer or cover.

### Points for consideration

– In the case of screeds laid on an elastic, flexible layer of insulating material all requirements must be met, such as their crushing, flexural and shearing strengths, thickness and hardness (compression resistance) as well as their shrinkage and surface evenness.

– Crushing and flexural strengths must be adequate to absorb the static and dynamic loads related to the compressibility of the insulating layer and the thickness of the screed, and to distribute these loads over the whole surface of the insulating layer without resulting in cracking. These various strengths are affected mainly by the type of aggregate used, by the water/cement ratio, and by compaction. The average strength of the screed should not fall below the following values: the prism crushing strength must be at least 22.5 N/mm$^2$, the prism flexural strength must be at least 4 N/mm$^2$ or 2.5 N/mm$^2$ during construction. The minimum required thickenss of the screed layer depends on the compression of the insulating layer, since this in turn affects the flexural stresses that occur. Attention should also be paid to irregularities in the substratum as well as to the fact that it may not be possible to ensure complete and uniform compaction of the lower layers of the screed, since these factors can reduce the effective load-bearing section.

– The shearing strength of a screed can play a particular role if the screed is subjected to shearing stresses as a result of a surface covering bonded directly to it. Moreover, the hardness of the screed surface can be particularly important, especially if it is covered with soft materials and is exposed to spot loads. In this case, minimum requirements are not specified. However, screeds which have sufficient flexural strength and with particularly good surface compaction provide sufficient resistance to the loads normally associated with residential buildings.

– Stresses (inherent stresses and compressive stresses) may be produced in the screed as a result of shrinkage; these stresses may exceed the low initial flexural strength of the screed and thus lead to cracks. In order to reduce this risk of cracking, large continuous screed surfaces are subdivided by means of shrinkage joints, so that no one screed surface exceeds 6 m in length. Since joints in screeds represent defects (see C 2.1.2 – Joints in screeds) efforts must be made to minimise the shrinkage of the screed and the shrinkage stresses which occur. These depend on the type of cement used, the water/cement ratio, and particularly the drying conditions (see C 1.1.5 – Construction and after-treatment).

Floors
Floating cement screeds

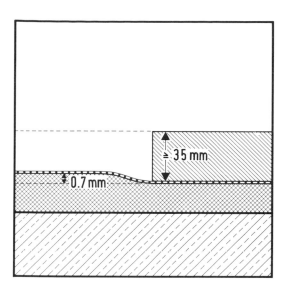

– The type and grade of aggregate used has a decisive influence on the strength of the screed. Irregular grading results in porous screed structures. Thus, aggregates with a grain size of up to 8 mm, and with a constant structure are suitable for use in screeds. The shape of the aggregate grains also has an effect on the workability and compaction characteristics of the screed; spherical grains and washed aggregate are the best.

– The strength and proportion of cement used have an effect on the strength, water requirement and workability of the screed mortar. In order to obtain an adequate screed, the quantity of cement used per m³ of finished mortar should be between 350 and a maximum of 400 kg.

– The water content of the mortar consists of the mixing water and the moisture contained in the aggregate. For complete hydration of the cement, about 40% wt. of water is necessary, based on cement content. However, as a rule it is necessary to use more water in order to ensure adequate workability and compaction characteristics. It should also be remembered that as the proportion of water increases, there is a clear reduction in final strength; at the same time, shrinkage increases. The water requirement can be reduced and the workability of the mortar improved by using various additives (admixes – follow manufacturer's instructions carefully). However, the influence of these additives on the final strength of the screed must be taken into account, e.g. the effect of admixes on screed strength.

### Recommendations for the avoidance of defects

● Care should be taken to ensure that the screed mortar is correctly and uniformly mixed. It should consist of a washed aggregate with a maximum grain size of 8 mm, with between 350 and a maximum of 400 kg of cement per m³. The water/cement ratio should not exceed 0.6 and the consistency should be between stiff and plastic. Additives can be used to improve workability. However, they must not have an adverse effect on the strength of the screed.

● The average strength of the screed should not fall below the following values: the prism crushing strength must be at least 22.5 N/mm², the prism flexural strength must be at least 4 N/mm², or the flexural strength of a screed sample taken from the structure must be at least 2.5 N/mm².

● The minimum thickness of the screed depends on the compression characteristics of the insulating layer. In the case of the typical loads found in residential buildings, and taking into account the compression characteristics of the insulating layer, it should be between 35 and 45 mm thick. (British Standard CP 204 Part 2 recommends a thickness of not less than 65 mm.)

● The continuous surface area of the screed should not exceed 30 m² and the maximum length should not exceed 6 m.

Floors
Floating cement screeds

Many different types of defect to floors could be attributed to incorrect construction of the screed or to incorrect after-treatment. The following types of defect occurred: both reticular cracks in the surface of the screed and cracks which penetrated the depth of the screed and which fractured large continuous screed surfaces, as well as breaks in the screed and lifting of surface finishes in areas subjected to frequent traffic.

In some cases, the screed was too thin in places, was full of cavities, was too weak, or its surface was fissured and uneven. Sound insulation was also reduced if there were acoustic bridges in the insulating layer.

**Points for consideration**

– In addition to correct composition and size (see C 1.1.4 – Material selection and size), the construction and after-treatment of the screed also have a decisive influence on ensuring that the screed will be successful.

– A sufficiently homogeneous mixture will be obtained merely if all the components are mixed for 3 min. in a power mixer. However, since the mortar can separate during transportation, it should be used as quickly as possible.

– Once laid, the layer of insulating material can become displaced or creased if walked or driven on (or when the screed mortar is tipped), thus producing acoustic bridges or weak points in the depth of the screed. This danger can be reduced if sufficiently wide boards are laid down, for traffic. It is best if the mortar is distributed over the screed which has already been laid.

– Levelling out the wet mortar does not provide sufficient compaction, nor does manual compaction, since the flexible insulating layer is springy. Alternatively, a surface vibrator distributes the cement lime evenly throughout the screed's depth and beds the aggregate grains tightly together, thus reducing the volume of cavities in the screed.

– Depending on the type of surface covering to be used, as well as on the intended use of the building, the surface of the screed must be level. Once the screed surfaces have been laid, they are smoothed to provide additional compaction and levelling of the surface. Boards or panels are suitable for distributing the weight of the floor layer when working. Nailed boards carry the risk of producing acoustic bridges, since they may damage the covering or insulating layers.

– The appearance of reticular shrinkage cracks can be avoided and adhesion of bonded surface finishes can be improved if the surface of the screed is levelled with a felt board (sponge) to remove any accumulation of cement lime on the surface. Once it has hardened, the surface can be levelled by buffing.

– Because of the large surface area of the screed, its thin structure and its damp consistency, all factors which promote evaporation – increased temperatures and movement of air, direct sunlight and low relative air humidity – result in rapid water loss. The drying speed of the screed affects both its strength development and the deformation and inherent stresses in the screed. Moreover, the strength of the screed will be severely reduced if the water which it requires for complete hydration is lost. Shrinkage contraction is also dependent on drying characteristics. If there is extensive shrinkage at the beginning, the shrinkage stresses which occur will be greater than the initially low strength of the screed and cracking will result.

Floors
Floating cement screeds

– Preventing drying out in the period immediately following laying, delays shrinkage processes and has a favourable effect on strength development. However, for this to be possible, all openings in the building must have been closed before the screed is laid and the screed must be moistened immediately after it is laid and then covered for at least seven days with an impervious PVC strip, e.g. polyethylene film. The use of synthetic resin additives in the screed mortar improves water retention characteristics and thus simplifies the after-treatment measures that are necessary.

– Since the strength of screeds increases slowly, there is a risk of damage caused by footprints (especially around the edges or at joints) if they are walked on too early. An interval of at least three days should be allowed before a screed is walked on for the purposes of inspection; however, as a rule, the earliest that a screed will have reached sufficient strength for it to be used as intended is three weeks. Suitable measures must therefore be adopted to prevent access to the screed.

### Recommendations for the avoidance of defects

● The cement screed mortar must be prepared immediately before use and must be homogeneously mixed in a power mixer for at least 3 minutes.

● Whilst the screed mortar is being transported and laid, care should be taken to ensure that the insulating and covering layers do not become displaced or damaged. They should be protected by boards to distribute any loads.

● The screed must be compacted effectively. Where possible, surface vibrators should be used.

● The compacted surface of the screed should be trowelled level over accurately adjusted guides and should then be smoothed. The tolerances specified should be observed according to the particular application involved. Levelling with a felt board (sponge) will roughen the surface.

● The finished cement screed must be kept damp for a period of at least 7 days and should then be kept from drying out for a further period of at least 14 days. Openings in the building (doors and windows) should therefore be closed before the screed is laid. Once it is complete, the screed should be wetted and covered for at least 7 days with a polyethylene film.

● Screed additives, such as synthetic resin admixes, can be used to improve the water retention capacity of the mortar, thus simplifying the necessary protection measures.

● Effective measures must be taken to prevent access to freshly laid screed surfaces.

Floors
Floating cement screeds

If floating screeds were laid on floor slabs (on the ground floor of buildings with no basement or in basements) that were in direct contact with the ground and which had no damp-proof course other than a layer of material to prevent capillary action, defects were found over the whole floor surface. The screed itself, and the insulating material beneath it were damp and – if the insulation was made of vegetable materials – was damaged by rot. Similarly, surface finishes also became damaged. This type of defect was also found in floor slabs which had been finished merely with water-proof screeds or sealing compounds.

In those cases where the damp-proofing measures took the form of bonded bitumen sealing courses or plastic films which were tighly connected to the horizontal sealing courses through the cross section of the walls, damp penetration damage was found only along the walls.

### Points for consideration

- Structures immediately adjacent to the ground are permanently exposed to moisture as a result of capillary action and diffusion if there are quantities of water in the soil, or in the presence of water under hydrostatic pressure. Even floor slabs which are separated from the ground are exposed to high moisture levels.

- The dryness requirements of the substratum for floating screeds are more exacting, partly because demanding uses will be made of them and partly because the floating screed, the insulating layer, and/or the floorcovering can become damaged by moisture.

- Fills to prevent capillary action, as well as water-proof screeds and sealing compounds do not adequately prevent the movement of moisture by capillary action or diffusion through the reinforced concrete slab. Damp-proof seals which consist of two layers of bituminous felt, or plastic films which are bonded over the whole of their surfaces offer good and safe protection against damp penetration, even when the demands for floor dryness are very exacting. However, this is conditional upon the provision of suitable measures at wall connections to prevent the ingress of moisture or water.

- Because of the continuously low temperatures of the ground, it is advisable in the case of heated rooms to increase the thermal insulation value of the floor structure by the use of additional thermal insulation layers.

- The thermal resistivity of floors in heated rooms should be $\geq 0.86$ m$^2$ K/W (coefficient of heat conductivity of k $\leq 0.97$ W/m$^3$ K). When making the calculation, only those layers within the seal should be taken into account.

Floors
Floating cement screeds

**Recommendations for the avoidance of defects**

● Floor surfaces in contact with the soil which are to be covered with a floating screed must have a damp-proof and capillary-tight seal beneath the layer of insulating material. This can be provided by two layers of sealing membrane bonded together over the whole of their surface, or by single layer heavy-duty plastic film with welded seams; these sealing courses should then be tightly joined to the horizontal wall damp-proof course.

In heated rooms, the thermal resistivity of the floor should be at least $0.86 \text{ m}^2 \text{ K/W}$ ($k \leq 0.97 \text{ W/m}^2 \text{ K}$). Care should also be taken to ensure that sound insulation standards are met.

● In rooms which have increased air humidity, insulating layers with a low level of vapour resistance should be covered with a vapour barrier. This can serve simultaneously as a separating layer.

Floors
Floating cement screeds

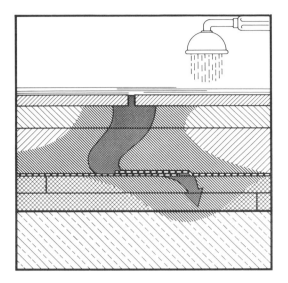

Rot was found in insulating layers made of vegetable fibres and in timber joists beneath bathrooms, shower rooms and utility rooms, and damp damage was found in the ceilings beneath these rooms if the floating floor structures were not provided with a water-proof sealing layer – or if joints in the surface of this sealing layer were incorrectly made. Moreover, if steel pipes, such as heating pipes, were installed in these floors without corrosion protection, these became severely corroded and in some cases leaked.

**Points for consideration**

– A cement screed – similarly to a tiled surface – is not made permanently water-proof merely by the inclusion of sealing agents. Hairline cracks between the tiles and the mortar caused by different rates of shrinkage cannot reliably be prevented, and a screed laid on a flexible substratum cannot be fully compacted.

– A conventional covering layer laid over the insulation material is not permanently moisture and water-proof, as its seams are not bonded. Similarly, junctions with pipes and wall connections are not water-proof.

– In damp rooms in residential buildings (e.g. bathrooms, shower rooms, utility rooms) it can be assumed that the floor will only be subject to limited quantities of water (splashes, cleaning water) for temporary periods. Therefore, it is a question of stopping moisture from penetrating the insulating layer and adjacent structures.

– In wet rooms (such as laundry rooms, private and public swimming baths, shower cubicles) the degree of exposure resulting from use and cleaning (with a hose) is intense; even the wall skirting is exposed to splash water. Moreover, if the surface of the floorcovering is not laid with a slope towards a floor drain, puddles can be expected to form. Similarly, water can accumulate on the sealing layer if this is not laid on a slope and is not connected to a drain. The sealing of wet rooms, including all connections, must therefore separate the floor surface finishing from all adjacent structural components in the form of a 'tank construction' and must ensure that no water is able to seep through the connections (see C 2.1.3: Connection of the seal to adjacent structures).

**Recommendations for the avoidance of defects**

● In damp rooms, in residential buildings, which are not heavily exposed to moisture, a single layer damp-proof course with bonded seams can be laid directly on top of the insulating layer. However, because of the particularly flexible insulating materials used, care should be taken to ensure that this damp-proof course does not become damaged when it is laid, and particularly when the screed is laid.

● However, if intense water exposure is expected, the damp-proof course must consist, for example, of two layers of sealing membrane bonded together over the whole of their surfaces and laid with a fall of 1.5%; they should also, like the floor surface, be linked to a floor drain. The fall should be obtained either by a sloping screed or by a compacted filling.

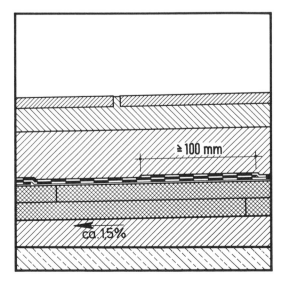

≧ 100 mm

ca. 1.5%

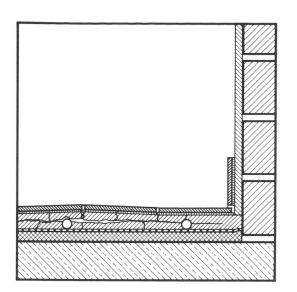

Floating floor screeds with built-in heating systems were repeatedly found to be defective: tiled floor finishes made of stone or ceramic material lifted and warped along the walls or at supporting points and cracks appeared in the screed and the tiling if large areas of the floor had been constructed in one area or if the tiling continued directly to adjacent structures. In many cases, this was associated with a reduction in impact sound insulation. Severe cracks which followed the path of heating pipes appeared if these pipes were embedded in the screed, thus severely reducing its thickness at various points.

If the underfloor heating was tested a few days after the screed was laid, the screed was too weak; it cracked and was easily worn when walked on. Moreover, tiled finishings had lifted over large areas.

Signs of corrosion appeared in damp rooms or swimming pools if the pipes did not have sufficient corrosion protection and if water was able to penetrate to the pipes.

**Points for consideration**

– The fluctuating temperatures (approx. 50 K) which occur in the screed when the heating is used, cause expansion. When it heats up, the screed – and the surface finish – expands. Thus, if there are not sufficient perimeter expansion joints around the edges of the screed, or where other structures make contact with the screed, compressive forces are produced in the screed and the surface finish and these can, in turn, cause the surface finish and the screed itself to warp. On cooling, the screed contracts – and with it the surface finish. Thus, if the continuous screed surfaces are too long, they can become highly elongated, or bent (where other structures project into the room), thus producing tensile stresses in the screed that can lead to cracks because of the brittle nature of mineral materials.

– Such cracks cannot be avoided by inserting reinforcement (reinforcing steel mesh). However, where the screed is sufficiently thick, and if they are correctly positioned in the area subject to stresses, they can increase the load-bearing capacity of the screed skin.

– Heating to a high temperature, such as when the heating system is tested, will cause rapid drying out of the screed. The strength development of the screed is interrupted because the water it requires for complete hydration is removed, and so the screed will have only a fraction of the strength it would have had if it had dried out slowly. Thus, the screed will not be able to absorb without damage the loads that will later be imposed on it when the tiling is laid and the building is in use.

– Moreover, a rapid loss in water will result in a correspondingly large amount of shrinkage in the screed mortar, and as a result of the latter's low initial strength, this will result in an increased number of shrinkage cracks in the finished screed.

– Joints in the form of trowel cuts are only suitable for the specific absorption of one-off contractions in the screed as a result of shrinkage. Expansion joints, which are designed to absorb the constant expansion and contraction of the screed, must be provided and must be constructed according to the anticipated expansion and the materials used, and must separate the screed together with the floorcovering (see C 2.1.2: Joints in screeds). Rigid connections, e.g. as a result of pipes or heating wires, or steel reinforcement laid in the screed, prevent the action of expansion joints and can result in damage to these elements in the area of joints. Similarly, acoustic bridges

Floors
Floating cement screeds

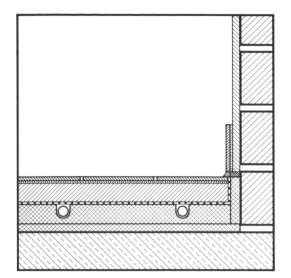

may occur at points of contact between the screed, or a solid floorcovering and the walls, which can considerably impair impact sound insulation.

– If water and oxygen come into contact with pipes made of ungalvanized steel, rapid corrosion will occur. This process is accelerated by the increased temperatures associated with underfloor heating systems and by the presence of corrosive components in the screed. This problem arises not only in damp rooms – bathrooms, utility rooms – but also wherever there is a risk of the screed surrounding the pipes being occasionally or repeatedly exposed to moisture, e.g. as a result of leaks in the pipework or because of cleaning water.

Water-proof seals are one way of protecting the floor structure in wet rooms from being exposed to damp (see C 1.1.7: Damp and wet rooms). A further way of reducing the risk of corrosion (even in living rooms) is to provide the pipes with sufficient all-round protection against corrosion.

**Recommendations for the avoidance of defects**

● The heating elements of underfloor heating systems should be laid in their own bed which should form a level and suitably strong basis for the screed. Suitable beds are, for example, layers of thermal insulation material with channels and screed layers of sufficient depth.

● The continuous floor area of a screed above underfloor heating systems should not exceed 20 m², and the maximum continuous length should not exceed 5 m.

● Suitable perimeter or expansion joints should be arranged along all edges and around supports and projections in the floor profile. The individual part surfaces formed in this way should be constructed as separate heating circuits.

● The thermal insulation layer beneath underfloor heating systems should have a thermal resistivity of $\geq$ 1.25 m² K/W.

● Underfloor heating systems should not be put into operation until the screed and mortar or adhesive layers of surface finishes have fully hardened and dried out. In the case of cement screeds, `at least 28 days should be allowed after laying. After this, the temperature should be raised gradually, in daily intervals of about 10 K.

● The pipes used for underfloor heating systems should be manufactured of corrosion resistant material or should be provided with lagging to protect them from corrosion.

Floors
Floating cement screeds

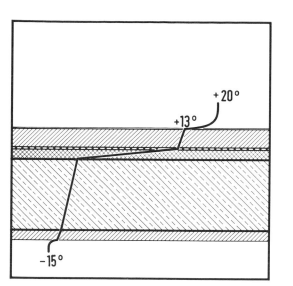

The following damage and defects were found in floors which were adjacent to the atmosphere or to non-heated internal rooms, or which were provided merely with impact sound insulation:

After a relatively long stay in the room, the floors were cold on the feet and a large amount of heating was necessary to warm the air in the room. In some cases, the floor finishes had been penetrated by damp and there was a musty smell, patches of mildew and condensation were prevalent if the rooms were not permanently heated, or if the only heating they received came from adjacent heated rooms. In the case of cantilevered floor slabs or those with voids below, condensation formed within the floor structure, for example on the surface of the reinforced concrete slab where the finished floors were on timber joists.

Similar defects occurred in floors in contact with the ground. These were discussed in C 1.1.6 – Floor slabs in contact with the ground.

### Points for consideration

– Floor slabs in contact with the outside air, such as slabs over passages, balconies etc., are external structures. They are exposed to the extreme temperature fluctuations of the external walls, especially in winter, and must therefore be constructed taking thermal insulation factors into account. Floor slabs above non-heated internal rooms, such as basements, also fulfil a thermal insulation function. However, because of the smaller temperature differences, the demands made on them are less.

– If the thermal resistivity of a floor structure is low because there is no insulation, or because there is only one thin impact sound insulation layer, the high temperature losses resulting from the prevailing temperature gradient cause high temperature losses, with the temperature of the floor surface remaining low.

– If the temperature of the floor surface falls below the dew point of the air inside the room, condensation will occur and cause damp penetration.

– Despite sufficiently high air temperatures, floors with a low surface temperature will feel cold to the feet after a relatively long period in the room and in extreme cases may constitute a health hazard. In order to obtain comfortable air temperatures, the thermal resistivity must be such that the internal surface temperature amounts to at least 17°C, with an air temperature of + 20°C. Even floor finishes with a low degree of heat dissipation (carpets) cannot compensate for the low thermal resistivity. They only improve comfort in the room for brief periods.

– If, as a result of the construction, and the materials used in the floor, water vapour can enter its cross-section in such large quantities that its saturation point is exceeded by a considerable amount, this water vapour will condense and in some cases cause damp damage in the floor structure. Structures particularly at risk are those which allow direct contact between warm internal room air and non-insulated, and therefore cold reinforced concrete slabs (dry floors on timber joists), or those which have thermal insulation layers with a low degree of diffusion resistance above the reinforced concrete slab, or which are covered with surface finishes with a low vapour barrier value (carpets).

– As a rule, the danger of harmful condensation in the cross-section of the floor structure is slight if, irrespective of the type of thermal insulation material used, the latter is fitted

Floors
Floating cement screeds

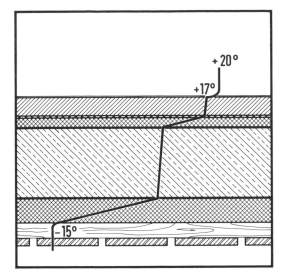

to the underside of the reinforced concrete slab. At the same time, this also solves the problem of levelling floor surfaces with and without thermal insulation on one storey.

– The following requirements are recommended in terms of thermal resistivity (or coefficient of thermal conductivity):
  – in floors above non-heated basement rooms $\geq$ 0.86 m² K/W (k $\leq$ 0.97 W/m² K);
  – in floors which are suspended in contact with open air: $\geq$ 1.72 m² K/W (k $\leq$ 0.49 W/m² K).

Alternatively, the thermal insulation recommendation requires the following thermal resistivity (coefficients of thermal conductivity):
  – in floors above non-heated basement rooms: $\geq$ 0.91 m² K/W (k $\leq$ 0.80 W/m² K);
  – in floors which are suspended in contact with open air: $\geq$ 1.88 m² K/W (k $\leq$ 0.45 W/m² K).

### Recommendations for the avoidance of defects

● The size of slabs above non-heated internal rooms (basements), or above ground level must be chosen bearing thermal insulation in mind.

● The thermal resistivity of slabs above non-heated basements must be at least 0.86 m² K/W (coefficient of thermal conductivity $\leq$ 0.97 W/m² K).

● The thermal resistivity of slabs which are in contact with the open air must be at least 1.72 m² K/W (coefficient of thermal conductivity $\leq$ 0.49 W/m² K).

● Where possible, the necessary layer of thermal insulation should be fitted to the underside of the ceiling slab. If, however, it is necessary for it to be fitted to the upper side, the thermal insulation materials used should have a high resistance to vapour. Thermal insulation materials which have a low resistance to vapour (e.g. glass wool) must have a vapour barrier on their upper surface in the form of laminated aluminium foil or similar.

## Problem: Bonded screeds

In residential buildings, bonded cement screeds are used predominantly in ancillary rooms, garages, storerooms and basements. They are also used as smoothing renders for base slabs which already meet the necessary requirements in terms of air-borne sound insulation – the impact sound insulation being provided by a flexible surface finish. Unlike floating screeds, bonded screeds play no part in improving the sound insulation or the thermal insulation of slabs.

Bonded screeds are used either as a substratum for surface finishes or are used as surface finishes for direct wear. In both cases the requirements differ in terms of evenness and hardness, strength and appearance of the surface. Unlike floating screeds laid on separating layers so that they 'float', bonded screeds form a strong bond with the substratum over the whole of their surface, so that deformation and expansion and contraction of the screed is practically impossible. This deformation is replaced by internal stresses, and more particularly by shearing stresses in the area of the bond. Thus, both the substratum – usually (reinforced) concrete slabs – and the screed must be able to withstand these special stresses.

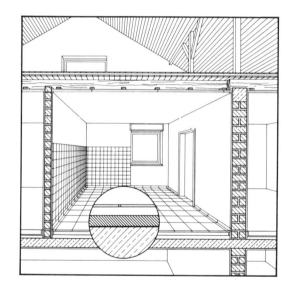

The way in which bonded screeds are constructed, and the stresses involved result in recurrent defects, which can be attributed to typical faults. The main areas of failure stem from the adhesive bond, which can be attributed to the unsuitability of the substratum and/or inadequate preparation of the substratum, to the incorrect choice of materials and make-up of the screed mortar, and to a failure to treat the freshly laid cement screed. These areas of structural failure are described and analysed below, and recommendations are made for the avoidance of defects.

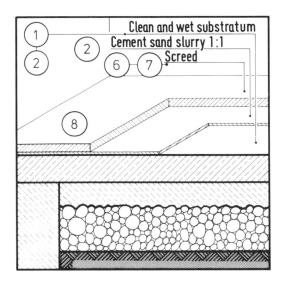

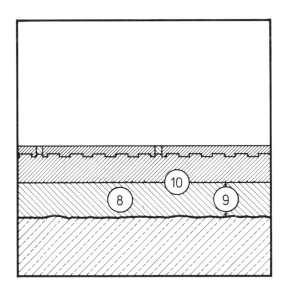

**1** The substratum for bonded screeds must be sufficiently rigid, crack-free, clean, level and textured (see C 1.2.2).

**2** Where possible, the screed should be laid on a concrete surface which has barely set (which is no more than 1 to 2 days old) and which is still damp (see C 1.2.2).

**3** If the substratum has already hardened and is dry, it must be wetted at least 24 hours before the screed is applied. Immediately before application of the screed, a cement–sand slurry (mixing ratio 1:1) should be brushed in (see C 1.2.2).

**4** If the substratum has already been treated with a surface finish such as a sealing compound, an adhesion agent such as a synthetic resin admix must be used (see C 1.2.2).

**5** Above movement joints in the substratum, as well as above joints between slabs, etc., a suitable expansion joint or a trowel cut joint must be made in the screed (see C 1.2.2).

**6** Attention should be paid to ensuring that the composition of the screed mortar is uniform and correct. It should then be homogeneously mixed in a power mixer for at least 3 minutes. It should consist of a washed aggregate (if necessary with suitable hardeners) with a maximum grain size of 8 mm and up to 350 kg of cement per $m^3$ of finished mortar. The water/cement ratio should not exceed 0.6 and its consistency should be between stiff and plastic (see C 1.2.3 and C 1.2.4).

**7** Additives or admixtures, such as synthetic resin can be used to improve the workability, adhesion and water retention capacity of the screed, as well as to reduce its modulus of elasticity (see C 1.2.3 and C 1.2.4).

**8** The quality of the screed must meet the minimum values specified in CP 204 Part 2 'In-situ Floor Finishes' (see C 1.2.3).

**9** The bonded screed must be at least 40 mm thick (see C 1.2.3).

**10** Once the moist screed has been dispersed over the prepared substratum, it should, where possible be compacted with a vibrator board or a surface vibrator. It should be trowelled level and smoothed with the aid of accurately adjusted gauges. The tolerances specified should be observed according to the particular application concerned. The surface can be roughened by trowelling with a felt board (sponge) (see C 1.2.4).

**11** The finished screed must be kept damp for a period of at least 7 days and should then be kept from drying out for a further period of at least 14 days. Openings in the building (doors, windows) should be closed wherever possible (see C 1.2.4).

**12** Effective measures must be taken to prevent access to newly laid screed surfaces (see C 1.2.4).

Floors
Bonded screeds

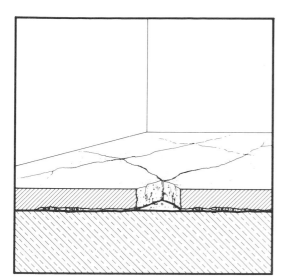

Defects in bonded screeds in the form of breaks and open cracks were particularly common if the intended all-over bond between the substratum and the screed had not been attained, so that there were in some places cavities beneath the screed. The substratum, which generally took the form of concrete slabs or reinforced concrete slabs which were in contact with the ground, were either soiled, too dry or, for example, treated with sealing compounds. Cracks also appeared in the screed above structural joints or cracks in the substratum.

**Points for consideration**

– In order to produce a screed which serves its purpose correctly and is free of defects, it is important to ensure a good bond between it and the substratum and to ensure that the substratum is free of defects and cracks.

–The strongest bond between the screed and the base slab is attained if the screed is applied to the slab when it has barely set, i.e. no more than 1 to 2 days after the slab has been concreted. However, this is frequently not possible because of difficulties in organising the building schedule.

– If the screed is to be applied at a later date to concrete which has already hardened and is dry, special measures are necessary where the substratum is concerned. If the strength of the concrete is considerably less than that of the screed, there is a risk of fracture in the concrete. Moreover, if the surface is brittle and dirty, or if it appears that it has been treated with a wax-type film, it will be impossible to produce a strong bond over the whole of its surface.

– An uneven substratum will result in considerable fluctuations in the thickness of the screed, which is itself 40 mm thick and may for example, result in cracking when it dries. In contrast, a rough surface will serve to promote a good bond.

– Wetting the substratum at least 24 hours before the screed is laid will prevent removal of the water which the bonded screed requires for complete hydration. If this is not done, the strength of the screed and the adhesive bond produced will be reduced, and weakened, respectively. Moreover, a cement–sand slurry brushed into the concrete immediately before the screed is applied will improve the key between the two layers. Suitable synthetic resins may also be used to promote good adhesion.

– Special care and precautions are required when bonded screeds are laid on damp-proof membranes, since the water-repellent action of the hardened compounds will impede the water-based cement mortar in producing a bond.

– The bonded screed, which is also to serve as a protective layer for the damp-proof membrane, can therefore only be applied after the roughened surface of the compound has been treated with an agent to promote adhesion.

– Screeds consist of relatively thin layers of cement mortar and they are therefore prone to cracking. Although cracks caused by inherent stresses in the screed (e.g. shrinking) and by loads imposed by use can be prevented by ensuring a solid bond with the substratum, there is no guarantee that the substratum, and thus the screed, will be free of cracks, especially in the case of thin floor slabs that are insufficiently reinforced. Similarly, joints between different materials and structural joints in the substratum represent likely sources of cracking.

Floors
Bonded screeds

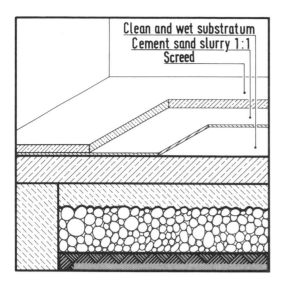

- The substratum for bonded screeds must be sufficiently rigid, crack-free, clean, level and textured.

- If the organisation of the building work allows, the screed should be laid on a concrete surface which has barely set (no more than 1 to 2 days old) and which is still damp.

- If the substratum has already hardened and is dry, it must be wetted at least 24 hours before the screed is applied. Immediately before application of the screed, a cement–sand slurry (mixing ratio 1:1) should be brushed in.

- If the substratum has already been treated with a surface finish such as a sealing compound, an adhesion agent such as a synthetic resin admix must be used.

- Above movement joints in the substratum, as well as above joints between slabs etc., a suitable expansion joint, or a trowel cut joint must be made in the screed (see C 2.1.2: Joints in screeds).

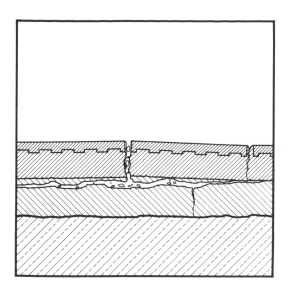

Screeds which were laid for direct traffic had a dusty, brittle surface which, in areas of frequent use, rapidly became worn or completely disintegrated. Tiles or slab coverings laid in a bed of mortar lifted together with the top layer of the screed. With both these types of surface covering fractures appeared along joints and at points where there was no bond with the substratum.

These defects were attributable to screed mortars of the wrong composition with an excessively high water/cement ratio, or unsuitable aggregates, or to fluctuating or inadequate screed thickness.

**Points for consideration**

– The composition of a screed mortar also has a decisive influence on the strength of a screed and on the production of a good bond over the whole of its surface. The thickness of the screed is also important.

– In a bonded screed, external loads are transmitted directly to the load-bearing reinforced concrete slab. If, however, there is no bond at certain points, or over the whole area of the screed, there will be cavities and so loads will be transmitted indirectly to the substratum via the screed skin which will deflect.

– Experience has shown that a screed at least 40 mm thick will be able to absorb without damage the sort of loads which occur in residential buildings, even if there are disorders in the bond at certain points. Moreover, as the thickness of the screed increases, it is able to equalise larger fluctuations in the substratum.

– The strength of the screed determines among other things its suitability for a particular load. Bonded screeds which are to serve as a backing for a floorcovering, for example, are exposed to lighter loads; in this case, a minimum cube crushing strength of 15 N/mm$^2$ is sufficient. However, bonded screeds which will themselves form the traffic surface are exposed to heavier spot loads and to greater wear. They should therefore have a minimum cube crushing strength of 25 N/mm$^2$.

– In addition to external loads, inherent stresses caused by deformations of the screed and of the substratum are of decisive importance in terms of shear stresses in the area of the bond. These shear stresses therefore increase as expansion or contraction occur (shrinkage, thermal expansion or contraction) and as the modulus of elasticity of the screed increases. Strength, shrinkage and modulus of elasticity are in turn affected by the characteristics of the aggregate used, the water/cement ratio, and by any admixtures that are used.

– Aggregates composed of washed, sufficiently rigid grains up to 8 mm in size will produce a screed with a sealed, tight structure. Spherical grains improve workability.

– The proportion and the strength of the cement have an effect on the water requirement and workability of the mortar. As cement content increases, and similarly as the water/cement ratio rises, the level of shrinkage also increases. The shrinkage of the screed is greater than that of the substratum merely because it contains larger proportions of binding agent and small grains; moreover, the substratum is generally also older than the screed and has therefore already undergone part of its shrinkage. Sufficient strength is attained if the cement content is equivalent to 350 kg per m$^3$ of finished mortar.

Floors
Bonded screeds

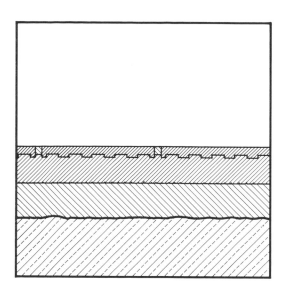

– The proportion of water used consists of the mixing water and the inherent moisture of the aggregate. For complete hydration of the cement, a water/cement ratio of 0.4 is necessary. As the water/cement ratio increases, workability is improved, but at the same time strength is reduced and shrinkage increased.

– Good workability is best achieved by the use of plasticisers. Additives made of synthetic resin also have a favourable effect, since they increase water retention capacity, improve substratum adhesion and reduce inherent stresses by lowering the modulus of elasticity.

– If it is necessary for the screed to have a sealing effect, special considerations must be taken into account. For further details on this see *Structural Failure in Residential Buildings, Volume 3 – Basements and adjoining land drainage.*

### Recommendations for the avoidance of defects

● Attention should be paid to ensuring that the composition of the screed mortar is uniform and correct: it should consist of a washed aggregate with a maximum grain size of 8 mm and up to 350 kg of cement per m³ of finished mortar. The water/cement ratio should not exceed 0.6 and its consistency should be betweeen stiff and plastic.

● Additives, e.g. plasticisers, can improve the workability of the mortar. In addition, admixtures such as synthetic resin can improve substratum adhesion and reduce inherent stresses. All additives should be checked to ensure that they do not have an adverse effect on the other properties of the screed.

● The strength of screeds exposed to normal loads must be as follows: the cube crushing strength of screeds to be covered with floor-coverings must be at least 15 N/mm², whilst the cube crushing strength of screeds to be used for direct traffic must be at least 25 N/mm².

● A bonded screed must be at least 40 mm thick.

Footprints and the imprints of trestles and stored materials, as well as structural weaknesses, were the result of the screed surfaces being used too soon after they had set. Individual open and reticular cracks occurred in the screed and brittle, rapidly worn screed surfaces were produced if the screeds were laid in summer conditions with high temperatures and with the doors and windows open. Irregularities and flaws in the screed surface could be seen through thin plastic floor-coverings. Slabs bonded to the screed surface tended to lift from the brittle screed.

**Points for consideration**

- In addition to correct preparation of the substratum (see C 1.2.2: Substratum), optimum mortar composition and sufficient screed thickness (see C 1.2.3: Material selection and size), compaction, surface treatment and drying conditions also play a major part in ensuring that the bonded screed will be free of cracks and will serve its purpose.

- Unlike a floating screed, a bonded screed can be well-compacted all over. With a stiff to soft mortar, it is best to use a board vibrator or a surface vibrator, since this will bed in the aggregate uniformly and densely and will produce a cavity-free structure.

- How even the surface of the bonded screed has to be depends on whether floorcoverings are to be laid on it, and what type of materials are to be used, as well as on what it is to be used for. Accurate levelling guides made of sectional boards are essential for producing a level surface and a screed of uniform thickness. If the screed is smoothed after it has been first trowelled, its surface finish will satisfy more exacting requirements – this also results in additional compaction of the surface area.

- Excessively long surface treatment can result in cement enrichment, which in turn leads to breaks in the screed or to shrinkage cracks. This risk can be overcome and the adhesion of any bonded floor-finishes laid subsequently can be improved if the surface of the screed is trowelled with a coarse sponge.

- In the case of screeds for direct traffic, the friction resistance of the surface can be improved by using particularly resistant aggregate, and by obtaining the best possible screed composition.

- Because the screed is relatively thin and yet has a large surface area, all factors which promote evaporation – increased temperatures and air movements, low relative air humidity – result in a rapid dissipation of water. Drying speed affects both the development of strength and inherent stresses in the screed. Shrinkage processes are delayed and strength development is improved if the screed is kept damp during the first 7 days, e.g. by spraying it with water and covering it with air-tight plastic sheets, and by avoiding conditions which would promote drying. The addition of synthetic resin admixes improves the water retention capacity of the screed mortar, thus simplifying after-treatment of the screed.

Floors
Bonded screeds

– If the newly laid screed is walked on, this will result in visible footprints and in structural failures. The screed should not be walked on for at least 3 days and it should not be fully used for at least 3 weeks. Effective measures must be taken to prevent access to the screed for this period of time.

### Recommendations for the avoidance of defects

● The correctly proportioned screed mortar – together with any special hard aggregates required – should be homogeneously mixed in a power mixer for at least three minutes.

● Once the damp screed has been dispersed over the prepared substratum, it should, where possible, be compacted with a vibrator board or a surface vibrator. It should be trowelled level and smoothed with the aid of accurately adjusted gauges. The tolerances specified should be observed according to the particular application concerned.

● The finished screed must be kept damp for a period of at least 7 days and should then be kept from drying out for a further period of at least 14 days. Openings in the building (doors, windows) should be closed wherever possible. The screed should be moistened after it has been completed and should be covered with plastic sheet.

● Effective measures must be taken to prevent access to newly laid screed surfaces.

# Problem: Floorcoverings

The floors of rooms in residential buildings are generally finished with a floorcovering. These floorcoverings consist mainly of stone and ceramic tiles, wooden parquet flooring, plastic and textile floorcoverings.

In addition to its appearance, the following factors are also of importance in chosing a type of floorcovering:

- sufficient resistance to wear, whilst being easy to clean

- the behaviour of the floorcovering in relation to climatic conditions and to conditions of use

- the floorcovering should be sufficiently warm to the feet and should, if necessary, provide sufficient impact sound insulation

- it should be suitable for the floor structure concerned.

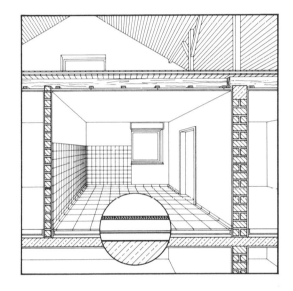

The survey revealed a wide range of defects in floors and floorcoverings. Almost all of these defects were to floors with tiled or parquet surfaces. However, commonly used plastic or textile floorcoverings were not undertaken in the survey, although they too are undoubtedly defective. The reason for this is certainly the fact that tiling and parquet flooring have a greater influence on the overall floor structure than the other types of floorcovering. In addition, new methods have been introduced over the last few years for laying these two types of finishes and this has meant changes in the conditions applied to their use.

The following section devotes itself entirely to a description of stone and ceramic tile floorcoverings and parquet floors, together with the defects they continually suffer. These defects are then analysed and recommendations are made for avoiding them.

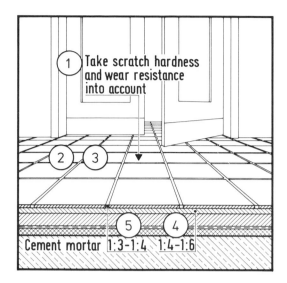

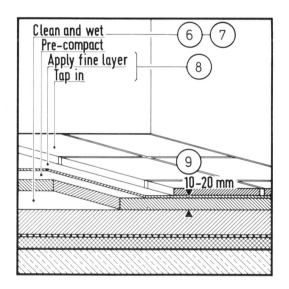

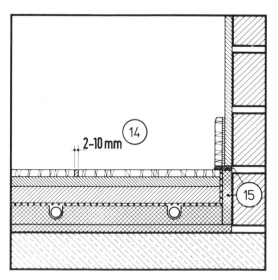

**1** The choice of tiled floorcovering depends on the loads to which the floor surface is to be exposed (see C 1.3.2).

**2** Fine ceramic tiles should meet the quality requirements of BS 1286. Where floor surfaces are exposed to heavy mechanical loads, unglazed quarry tiles or dense brick paviours should be used (see C 1.3.2).

**3** Natural flagstones for floorcoverings must be laid in a mortar bed which is generally no thinner than 15 mm (see C 1.3.2).

**4** The mortar used for laying clay tiles should be a cement mortar with a mixing ratio of 1:4 to 1:6 using sand with a grain size of 0–3 mm. Gauged mortar from mortar group II can be used for laying clay tiles inside buildings (see C 1.3.2).

**5** The grouting mortar should be a cement mortar with a mixing ratio of between 1:3 and 1:4, with a maximum sand grain size of about 1/3 of the width of the joint (see C 1.3.2).

**6** The substratum for tile coverings must be free of loose particles and impurities. Its level must not fluctuate by more than 5 mm in 0.1 mm when the tiles are laid in a mortar bed, or by more than 2 mm in 0.1 m when the tiles are laid using the thin bed method (see C 1.3.3).

**7** When laying tiles in a mortar bed, both the backs of the tiles and the dry substratum should be wetted (see C 1.3.3).

**8** After the mortar has been pre-compacted and trowelled, a fine layer (cement: fine sand 1:1 to 1:2) should be applied to the mortar if it is still damp before the tiles are laid and tapped in. If the consistency of the mortar is still plastic, it is merely necessary to powder the surface with cement (see C 1.3.3).

**9** After the tiles have been laid, the mortar bed should be at least 10 mm thick; however, it is better if it is between 15 and 20 mm thick (see C 1.3.3).

**10** If the bedding mortar – in the form of a steel reinforced 'screed-type' mortar bed – is to be laid direct on the foil-covered insulating layer, the application of two layers and careful compaction of the 40–50 mm thick mortar bed will ensure that its strength is similar to that of a screed (see C 1.3.3).

**11** If tiles are laid using the thin bed method, the substratum must be dry. It should be primed with a synthetic resin solution. The adhesive should be applied firmly and then trowelled up to provide a bed of adhesive at least 3 mm thick (see C 1.3.3).

**12** The floor area treated with adhesive in one operation should not be so large that the adhesive has started to set before the tiles are laid. Moreover, when the tiles are laid, they should be pressed firmly into the adhesive bed to ensure that about 80% of the backs is wetted (see C 1.3.3).

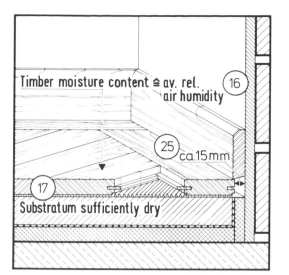

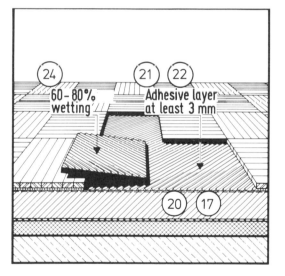

**13** Spread adhesives should not be used for surfaces which are exposed to moisture (see C 1.3.3).

**14** The width of the joints between the tiles should as a rule be no less than 2 mm and no more than 10 mm. Joints between tiles should be completely filled with grouting mortar, so that the grout is continuous with the surface of the tiles. It is particularly advisable to use cement mortars with suitably graded aggregates, or epoxy resin grouting compounds, in rooms that are exposed to moisture (see C 1.3.4).

**15** Around the perimeter of tiled floorcoverings and in the surface of the tiling itself, expansion joints should be made which match the edge joints and expansion joints used in the screed. These joints should be covered and filled with an elastic and water-proof material (see C 1.3.4).

**16** The moisture content of parquet flooring at the time when it is laid should be equivalent to the average moisture value. This should be checked by measurement (see C 1.3.5).

**17** Parquet flooring can only be laid on a sufficiently dry floor surface. The moisture content of a cement screed must therefore be no more than, for example, 1.5 times the practical moisture content. This should be checked by testing samples from the cross-section of the screed (see C 1.3.5).

**18** The whole surface area of floor slabs which are in contact with the ground, and including all connections (such as to the horizontal damp course of the walls) must be covered with a water- and capillary-proof sealing layer (see C 1.3.5)

**19** The thresholds of full-length windows or doors leading to roof terraces or balconies must be constructed in such a way that it is impossible for water to enter (see C 1.3.5).

**20** The substratum for parquet tiles must be sufficiently dry and solid (see C 1.3.6).

**21** The adhesive must meet the requirements of Code of Practice 201: 'Timber Flooring' (see C 1.3.6).

**22** The adhesive should first be firmly applied to the primed surface with a float. It should then be combed with a drag to produce an even layer of adhesive at least 3 mm thick (C 1.3.6).

**23** The area covered with adhesive in one operation should be such that the parquet can be laid before the adhesive has begun to set (see C 1.3.6).

**24** The individual sections of flooring should be pressed into the adhesive firmly and laid tightly together in order to ensure that about 80% of the back of each section is wetted (see C 1.3.6).

**25** Parquet floorcoverings which are laid on an adhesive bed, or which take the form of a floating structure, must be provided with expansion joints above joints in the substratum as well as at all connections, such as junctions with walls. They should also have expansion joints at suitable distances throughout their surface area. In general, the distance between joints should not exceed 6 m, or 5 m in the case of underfloor heating. The joints should be at least 15 mm wide (see C 1.3.6).

**26** Once the parquet adhesive has set, the floor surface should be rubbed down evenly and sealed in accordance with the room's intended use in order to provide a smooth and sufficiently wear-resistant surface.

Floors
Floorcoverings

Signs of serious wear – worn and scratched surfaces – were found in unglazed clay tiles, glazed fine ceramic tiles and natural flagstones or slates which were subjected to frequent traffic. In addition, the glaze of tiles was destroyed and natural flagstones began to flake if the surface was treated with acid cleaning agents or exposed to salt solutions.

The edges of tiles became damaged if joints were not completely grouted or if they were grouted with low-strength mortars. Large format tiles were particularly prone to damage and breakage if they were laid in a mortar bed which was not sufficiently strong.

Surface irregularities proved to be a nuisance when walked on or during cleaning operations, especially in the case of brick paviours.

**Points for consideration**

- Tiling on floors is subjected to much greater mechanical loads than wallcoverings. The suitability of a tile for use as a floorcovering therefore depends, above all, on the reaction of its surface to scratching and abrasion, as well as on the behaviour of the tile, after it has been laid, in relation to pressure.

- The scratch resistance of a tile depends on the scratch hardness of its surface. Clay tiles for flooring must be in accordance with BS 1286 for their quality and finish, dimensions and tolerances. Type A are usually known as floor quarries and Type B as floor tiles. Terrazzo tiles must be in accordance with BS 4131 for their quality and laying procedure, and specific recommendations are given for their maintenance. Thus, only terrazzo and quarry tiles and stone flags are recommended for use in entrances and corridors with heavy traffic wear.

- Bearing in mind the loads normally expected in residential buildings, tiles do not normally become damaged because their crushing strength is exceeded, but because they do not have sufficient flexural strength to withstand the loads imposed on them. Flexural stresses can be placed on an individual tile if there is a void beneath it or if the tile is laid in a flexible, low strength mortar bed; moreover, these flexural stresses increase as the length of the individual tiles increases. The size of the internal stresses produced when this occurs depends on the thickness of the tile. Thus, the reaction of a tiled floor to pressure loads is determined by the thickness of a tile made of a particular material in relation to its length, by the quality of the adhesive bond and the strength of the mortar bed.

- Thus, for example, according to BS 1286, glazed mosaic tiles 20 × 20 mm are 5 mm thick, whilst those measuring 250 × 250 mm are 30 mm thick for type A, and 150 × 150 mm type B, 12 mm. A mortar bed at least 15 mm thick is recommended for clay tiles.

- To achieve a good strength mortar bed specify cement mortar for ceramic tiles, gauged mortars from mortar group II for floor quarry tiles inside buildings (see C 1.3.3 for further details about mortar beds).

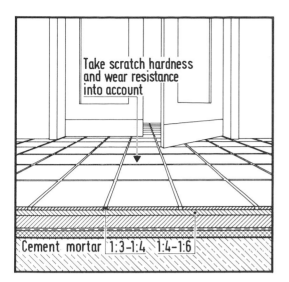

Take scratch hardness and wear resistance into account

Cement mortar 1:3-1:4 \ 1:4-1:6

- If a high-strength grouting mortar is well compacted and inserted so that it is level with the finished tiled surface, the risk of the tile edges breaking is slight. This also reduces shearing stresses at the edges of the tiles as a result of the differences in the expansion of the tiles and the substratum. Cement mortars are therefore particularly suitable as the grouting material (see C 1.3.4 for further details about joint width and application of grouting material).

- In entrances which are in direct contact with the street (e.g. front doors, entrance halls) solutions of thawing salt can come into contact with tiled surfaces inside the building, and like acidic-cleaning agents or household chemicals, they can attack unglazed tiles which are not acid resistant, as well as most types of natural stone. Fine ceramic tiles and cement grouting mortars are resistant to most chemicals found in residential buildings.

- Floor quarry tiles in particular – the raw materials of which contain a high proportion of water prior to cutting and baking – can deviate severely from the requirements of BS in terms of surface flatness (maximum permissible deviation $\pm$ 5% in relation to edge length) and can thus reduce the usefulness of the floorcovering.

### Recommendations for the avoidance of defects

● Floor tiles should be selected in accordance with anticipated surface loads.

● Fine ceramic tiles should meet the quality requirements of BS 1286, especially in terms of flexural strength, scratching hardness, surface wear, resistance to household chemicals and acids and in terms of surface flatness. Where floor surfaces are exposed to heavy mechanical loads (e.g. entrances, hallways) unglazed quarry tiles ('tiles with a low water absorption capacity') should be used.

● Natural flagstone tiles for floorcoverings should not be thinner than 15 mm when laid in a mortar bed. Materials which are sensitive to thawing salt should not be used in entrance areas.

● The mortar used for laying clay tiles should be a cement mortar with a mixing ratio of 1:4 to 1:6, using sand with a grain size of 0–3 mm. Gauged mortar from mortar group II can be used for laying clay tiles inside buildings.

● The grouting mortar should be a cement mortar with a mixing ratio of between 1:3 and 1:4 with a maximum sand grain size of about 1/3 of the width of the joint.

Floors
Surface coverings

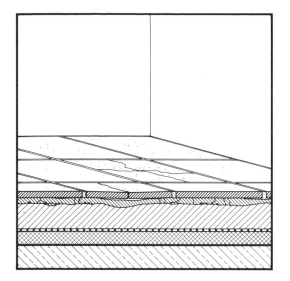

If the mortar bed for tiling was not sufficiently strong, or if there was insufficient adhesion between the tiles and the mortar bed because of errors when the tiles were laid, individual tiles parted and crushed when subjected to impact loads. Tiles spalled from large areas if the mortar bed had been applied to a sandy or soiled screed or if a thin bed adhesive had been applied to a screed which was still damp. Defects were particularly common in tiling laid over underfloor heating systems.

**Points for consideration**

– In residential buildings floorcoverings (screed and surface finish) are generally laid on low rigidity materials which provide impact sound insulation and in some cases thermal insulation. However, if, on the one hand, a firm all-over bond is not produced between the various layers placed on top of the insulating material and if provision is not made to allow the surface coverings as much freedom of movement as possible in relation to the insulating layer, then the differential deformation between the substratum and the floorcoverings as a result of deflection of the floor slab, shrinkage of the screed and thermal expansion and contraction of the floorcovering, can cause tiles to part from the substratum or to warp. Similarly, voids beneath tiles, or tiles that have been laid on flexible bases may be subjected to such severe flexural stresses as a result of direct traffic loads that they crack.

– The adhesion and firm all-over contact of the tile with the substratum must be achieved by means of a layer of mortar or adhesive and is thus dependent on the condition of the substratum surface, the properties of the mortar, or adhesive, used and on the bond produced between the mortar/adhesive and the tiles and the substratum.

– A mortar bed can generally only provide reliable adhesion with the substratum – generally a screed surface – if the latter is free of loose particles and impurities, and if it is neither so wet that no cement lime from the newly applied mortar can penetrate its surface nor so dry that the water which the mortar needs for complete hydration is removed from it.

– The requirements for tiles laid by the thin bed method are particularly stringent, both in terms of the strength, cleanliness and lowest possible water content of the substratum, and in terms of its evenness, since the thin bed is much less able to accommodate limited adhesion defects and irregularities than a mortar bed which is at least 10 mm thick (see C 1.1.4 for further details of screed construction).

– The properties of the mortar bed depend both on the materials used (see C 1.3.2) and on the thickness and compaction of the mortar after it has been applied.

– Careful compaction of the laying mortar takes on a particular significance if it is to be laid as a 'screed-type mortar' directly on a layer of insulating material which has been covered with a sheet of foil material, heavy polythene or felt. The reinforced mortar bed which has been laid in this way has also to fulfil the load distributing function of the screed, whilst totally enclosing the inserted reinforcing steel mesh in order to protect it against corrosion. Even where the mortar is laid and compacted in two layers, this type of construction remains much more susceptible to defects than tiling laid on an independent screed.

– Complete wetting of the backs of the tiles with mortar is improved if the tiles are tapped into the pre-compacted and trowelled mortar bed.

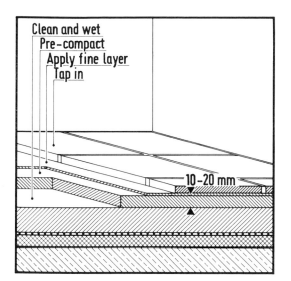

– A synthetic resin primer applied to the substratum will act as an adhesion bridge with paste-type adhesives and will ensure that the water which the thin bed cement mortar needs to hydrate will not be removed from it. Solvent-bearing primers cannot penetrate surfaces whose capillary pores are filled with water.

– The adhesive must be applied sufficiently thickly to ensure that an all-over bond is produced with the backs of the tiles, even in the event of slight irregularities in the substratum and in the case of tiles with profiled backs. The mortar bed is given an even thickness by combing it with a special trowel.

– A skin can form on the surface of thin bed adhesives as early as 10 minutes after application; in this state their adhesive properties are lost.

– Spread adhesives have less resistance to moisture than cement and epoxy resin adhesives.

### Recommendations for the avoidance of defects

● The substratum for tile coverings must be free of loose particles and impurities. Its level must not fluctuate by more than 5 mm in 0.1 m when the tiles are laid in a mortar bed, or by more than 2 mm in 0.1 m when the tiles are laid using the thin bed method.

● When laying tiles in a mortar bed, both the backs of the tiles and the dry substratum should be wetted.

● After the mortar has been pre-compacted and trowelled, a fine layer (cement: fine sand 1:1 to 1:2) should be applied to the mortar if it is still damp before the tiles are laid and tapped in. If the consistency of the mortar is still plastic, it is merely necessary to powder the surface with cement.

● If the bedding mortar – in the form of a steel reinforced 'screed-type' mortar bed – is to be laid direct on the foil covered insulating layer, the application of two layers and careful compaction of the 40–50 mm thick mortar bed will ensure that its strength is similar to that of a screed.

● If the tiles are laid using the thin bed method, the substratum must be dry (max moisture content in the case of a cement screed 2.5–3.0% wt. It should be primed with a synthetic resin solution.

● The adhesive should be applied firmly and then trowelled up with a drag to provide a bed at least 3 mm thick.

The floor area treated with adhesive in one operation should not be so large that the adhesive has started to set before the tiles are laid.

● When the tiles are laid, they should be pressed into the bed of adhesive (or even tapped in) sufficiently firmly to ensure that about 80% of the back of each tile is wetted.

Spread adhesives should not be used for surfaces which are exposed to moisture.

Floors
Floorcoverings

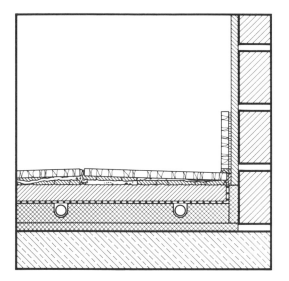

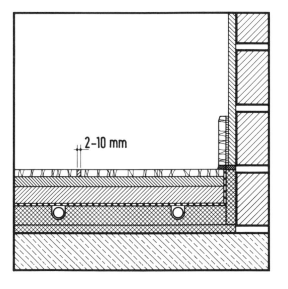

2-10 mm

Tiles cracked, or their edges lifted, if the joints between them were not fully grouted, or if the grouting mortar used was of low strength, or if the tiles were laid so that their edges ground against one another. Tiling laid above underfloor heating systems was particularly prone to lift and crack if there were no expansion joints where the tiling met the walls. In these cases, the floor slab also had defective impact sound insulation.

## Points for consideration

– If the mortar used in the joints between the tiles is of high strength, well compacted, and inserted so that it is level with the tile surface, the shearing stresses at the edges of the tiles as a result of the difference between the deformation of the tiles and the substratum are reduced.

– A joint less than 2 mm wide is very difficult to grout, whilst in joints over 10 mm wide, visible shrinkage cracks can appear in the mortar since it is rich in binding agents. These joints also serve to offset fluctuations in the size of the tiles; these variations can, for example, be greater in the case of quarry tiles compared with clay tiles. Moreover, in the case of air-tight floor finishes, moisture is dissipated from the mortar bed via the joint – wider joints are therefore advisable in the case of denser and larger tiles.

– In wet rooms, the floor must be sealed by means of damp-resisting courses (see C 1.1.7), since it is not possible to prevent completely leaks (hairline cracks) in the network of joints as a result of shrinkage of the grouting mortar, or deformation of the tiling (even when using water-proof grouting materials). Cement mortars mixed with correctly graded aggregates suffer relatively low shrinkage. Epoxy resin grouting compounds provide largely water-proof joints because of their capacity for expansion and their good adhesion.

– Tiling laid on a floating screed should be able to follow freely any deformation of the latter and should, for reasons of impact sound insulation, have no contact with adjacent structures. Thus, full-width expansion joints must be made in the tiling above all perimeter joints and other expansion joints in the screed substratum (see C 2.1.2–2.1.4).

## Recommendations for the avoidance of defects

● The width of the joints between the tiles should as a rule be no less than 2 mm and no more than 10 mm, the smaller size being used for small-sized tiles or mosaics, and tiles with small dimensional variations. Joints between tiles should be completely filled with grouting mortar so that the grout is continuous with the surface of the tiles. It is particularly advisable to use cement mortars with suitably graded aggregates (composition: see C 1.3.2) or epoxy resin grouting compounds in rooms which are exposed to moisture.

● Around the perimeter of the tiled floorcovering and in the surface of the tiling itself, expansion joints should be made which match with the edge joints and expansion joints used in the screed (see C 2.1.2–2.1.4). These joints should be covered according to their position and the mechanical and moisture loads to which they are exposed (e.g. by means of skirting tiles at wall connections) and should be filled with an elastic and water-proof material (e.g. by means of grouting compounds in floor surfaces in damp rooms which are not exposed to heavy mechanical loads) or by means of metal edging strips (in the case of surfaces exposed to heavy mechanical loads).

Floors
Floorcoverings

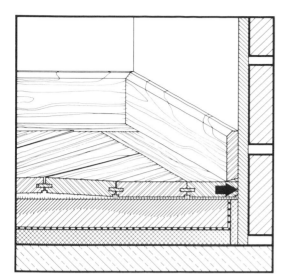

The defects found in parquet flooring could be attributed mainly to harmful effects from the substratum. These defects included discoloration, warping or cracking at points exposed to increased levels of moisture, for example over solid slabs or screeds which had not dried out fully, or above slabs which were in contact with the ground and which were inadequately sealed, as well as in the area of balcony and terrace doors. Further defects appeared in parquet blocks laid above underfloor heating systems, where open joints or cracks and – in the case of no perimeter joints – lifting of the wood blocks took place.

**Points for consideration**

– Of all the materials used in building, wood has the greatest capacity for swelling and shrinking. Thus, the moisture content of the wood at the time of installation, and after it has been installed, are important values which have a deciding influence on the way in which the parquet floor is constructed and laid.

– Because the cellular and fibrous structure of wood runs along the axis of the trunk, the swelling and shrinkage of wood when it absorbs or releases water – between complete dryness (kiln dry) and fibre saturation (about 30 % wt. – varies according to direction; that is, it is anisotropic. Whilst axial 'expansion' (across the grain) is practically negligible, the tangential 'expansion' (along the grain) of domestically grown timber is about twice as great as the 'expansion' which occurs in a radial direction. Moreover, the degree of shrinkage/swelling increases as the gross density of the wood increases. Thus, a heavy wood will shrink and swell more than a light one.

– The anticipated linear expansion of parquet flooring after it has been installed depends on the moisture content of the wood at the time of installation, based on its practical moisture content in situ. Alternatively, however, its practical moisture content is not constant, but is subject to considerable fluctuations, depending on variations in relative air humidity, among other things. Other influencing factors are the type and degree of room heating, the time of year, and, where applicable, direct sunlight.

– BS 1187 'Wood Blocks for Floors' specifies for domestic timbers at the time of delivery a moisture content of between 6 and 15 % wt. depending on the type of heating to be used in the house. The permanent moisture content of parquet flooring fluctuates between about 6 % wt. (in winter, with central heating, $\varphi_i = 30\%$) and 12–15 % wt. (in summer, with no air conditioning, $\varphi_i - 70$–80%). Oak parquet flooring will thus swell and shrink by up to about 26 mm/m over the course of the year if these movements are not prevented.

– Complete sealing of the parquet will retard the exchange of moisture. Thus, the effect of brief peak 'loads' on the moisture content of the wood can be reduced.

– In the case of parquet flooring laid on underfloor heating systems, the wood will dry out even more when the heating is on, so that swelling and shrinkage are correspondingly greater.

– If the parquet flooring is stored for a relatively long time in the room in which it is to be laid, this alone is no guarantee that its moisture content will correspond to the average permanent moisture level of the room. In any case, the moisture content of the room should be checked by measurement.

131

Floors
Floorcoverings

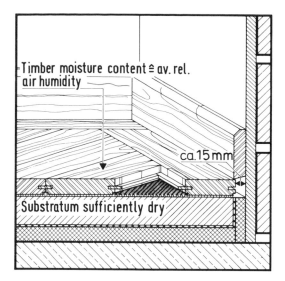

- The moisture content of the parquet flooring can be sharply increased after laying if the substratum (screed or solid slab) still has a high moisture content. As structures become thicker, reinforced steel structures and screeds take longer to dry out (depending on conditions in the room). This must be taken into account in deciding when to lay the parquet flooring in new buildings. The moisture content of the substratum when the floor is laid is of no importance if it does not exceed the practical moisture content to any major extent; in the case of a cement screed with a balanced moisture content of 1 to 2 wt. %, it should not exceed 2.5–3.0 wt. %.

- The moisture content of the parquet flooring may be increased at a later stage if moisture can reach the flooring after it has been laid, for example soil moisture where no damp resisting measures are provided, or where they do not prevent capillary action (use of a sealing compound) (see C 1.1.6: Floor slabs in contact with the ground), as well as because of precipitation in the area of balcony and terrace thresholds which are incorrectly constructed (see C 1.1.7: Damp and wet rooms).

- In addition, the parquet flooring can also become saturated with damp as a result of long term exposure to water – e.g. as a result of a leaking heating pipe, or as a result of cleaning water where the surface is inadequately sealed.

### Recommendations for the avoidance of defects

● The moisture content of parquet flooring at the time when it is laid should be equivalent to the average moisture value. This should be checked by measurement.

● Parquet flooring can only be laid on a sufficiently dry floor surface. The moisture content of a cement screed must therefore be no more than, for example, 1.5 times the practical moisture content. This should be checked by testing samples from the cross-section of the screed.

● The whole surface area of floor slabs that are in contact with the ground, and including all connections (such as to the horizontal damp course of the walls) must be covered with a water- and capillary-proof sealing layer.

● The thresholds of full-length windows or doors leading to roof terraces or balconies must be constructed in such a way that it is impossible for water to enter (see C 2.1.5: Threshold structure between indoor and outdoor areas).

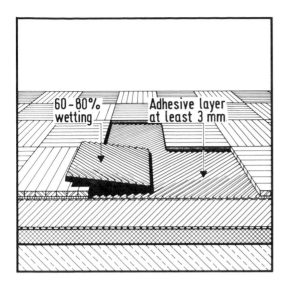

Some of the defects in parquet floors could be attributed to the use of the wrong materials and to incorrect laying. The defects concerned were: the parting of large areas of parquet flooring – and particularly basket weave pattern parquet – laid on screeds, and voids beneath individual parquet sections if, for example, the adhesive bond with the screed, filler compound or parquet blocks was not good, if the adhesive layer was too thin, or if there were irregularities in the substratum. In some cases, there were open joints in the flooring, irregular surfaces and signs of severe wear.

**Points for consideration**

- BS 1187 'Wood Blocks for Floors' describes the various types of parquet flooring normally used, in terms of their appearance, sizes and quality standards and a selection of timbers to be used.

- Depending on the type of parquet, the material and nature of the substratum and the use which is to be made of the room, parquet flooring can be nailed to battens or bonded to felt parquet backings or screeds. It can also take the form of a floating construction. Because of the particularly frequent defects which were encountered, space here will be devoted to discussing in greater detail the adhesion of parquet flooring, taking mosaic parquet as an example. Mosaic parquet consists of small pieces of wooden parquet whose edges have been machined smooth (without tongue and groove); freedom from defects thus depends on good adhesion to the substratum. Thus, the requirements are very strict in terms of substratum condition (evenness, dryness, strength and freedom from cracks and flaws) and the adhesive bond produced.

- To obtain a bond with sufficient shearing strength, it is important to obtain good adhesion of the adhesive to the substratum and the parquet.

- Smoothing compounds, which are used, for example, because of excessive irregularities in the screed surface, form a 'separating layer' between the adhesive and the screed if they are not strong enough, or if they do not provide sufficient adhesion with the screed. The hard plastic nature of the adhesive requires that a sufficiently thick film must be applied; this will, however, also permit harmless equalisation of expansion and stresses. Moreover, if thicker layers of adhesive are used it is also possible to smooth out unavoidable irregularities in the substratum.

- Primers which are compatible with the adhesive used will improve its adhesion to the substratum; solvent-based primers are able to penetrate deeper into the surface of the substratum, thus 'improving' its surface condition.

- Application of the adhesive with a float and subsequent combing with a special trowel will improve adhesion to the substratum and will ensure that the ridges of adhesive have a uniform depth of at least 3 mm. Sufficient wetting of the adhesive on the back of each section of parquet (60–80%) is most easily obtained if the parquet sections are pressed into the adhesive before it has begun to set (no film formation), tapped home and laid tightly together.

- Where parquet flooring is laid so as to float on a layer of insulating material, for example in the form of large panels of ready-made parquet bonded to plywood sheets, the demands made in terms of substratum condition are less stringent. Moreover, if by sufficient subdivision of the surface area (max. dimensions 5 m) and by using movement joints in the parquet surface and around the edges, the floor

133

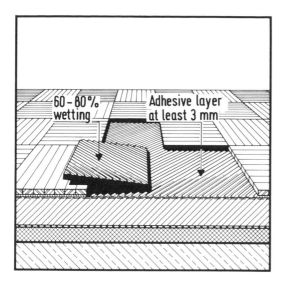

is able to expand and contract freely, this type of construction can also be used on floors with underfloor heating.

– The marked swelling and shrinkage characteristics of wood and the unavoidable fluctuations in air humidity in the room (see C 1.3.5: Parquet flooring – Substratum and moisture content) result in linear expansion and contraction of the parquet flooring after it has been laid. Moreover, if the flooring is unable to follow these movements, for example as a result of adjacent walls, the parquet floor may lift.

– The abrasive wear to which floors are exposed is variable. It is heaviest in the area of doors, staircases and hallways, etc., where static and moving loads (caused by friction, or rotary or sliding actions in different directions and at varying speeds) are frequent.

– As its gross density increases, the wear resistance of wood also increases. There are thus major differences between the various types of wood; however, there may also be differences in wood of the same type, because of various growth characteristics.

– When wear begins, the sealing products and protective films of sealers and polishes are first destroyed, then the wood is attacked directly. Wear is also increased if the floor is at the same time exposed to moisture (near a front door) since the wear resistance of wood falls as its moisture content increases. Wear resistance can be improved by using more resistant types of wood and more durable sealing products at points exposed to increased levels of traffic and by particularly intensive after-care.

## Recommendations for the avoidance of defects

● The substratum for mosaic parquet flooring must be sufficiently level, dry and solid. Any smoothing compound which has been applied must also be of sufficient strength and must have good adhesion to the screed.

● The adhesive must meet the requirements of Code of Practice 201 'Timber Flooring'.

● The adhesive should first be firmly applied to the primed surface with a float. It should then be combed with a drag to produce an even layer of adhesive at least 3 mm thick.

● The area covered with adhesive in one operation should be such that the parquet can be laid before the adhesive has begun to set.

● The individual sections of flooring should be pressed into the adhesive firmly and laid tightly together in order to ensure that between 60 and 80% of the back of each section is wetted. During laying, random checks on this should be made by lifting individual tiles.

● Parquet floorcoverings which are laid on an adhesive bed, or which take the form of a floating structure, must be provided with expansion joints above joints in the substratum as well as at all connections, such as junctions with walls. They should also have expansion joints at suitable distances throughout their surface area. In general, the distance between joints should not exceed 6 m, or 5 m in the case of underfloor heating. The joints should be at least 15 mm wide.

● Once the parquet adhesive has set, the floor surface should be rubbed down evenly and sealed in accordance with the room's intended use, in order to provide a smooth and sufficiently wear-resistant surface.

## Floating cement screeds

– Specialist books and guidelines

Forschungsgemeinschaft Bauen und Wohnen (Hrsg.): Estriche im Hochbau, Heft 80. Verlagsgesellschaft Rudolf Müller, Köln 1967.
Gösele, K.; Schüle, W.: Schall, Wärme, Feuchtigkeit. 4. Auflage, Bauverlag, Wiesbaden und Berlin 1977.
Henn, W.: Fußböden. Verlag Callwey, München 1964.
Schild, E.; Casselmann, H. F.; Dahmen, G.; Pohlenz, R.: Bauphysik, Planung und Anwendung. Vieweg Verlagsgesellschaft, Braunschweig 1977.
Schütze, W.: Der schwimmende Estrich, 4. Auflage. Bauverlag, Wiesbaden und Berlin 1974.
Schütze, W.: Estrichmängel – Entstehen, Vermeiden, Beseitigen. Bauverlag, Wiesbaden und Berlin, 1971.
DIN 4109, Blatt 2: Schallschutz im Hochbau; Anforderungen, September 1962.
DIN 4109, Blatt 4: Schallschutz im Hochbau; Schwimmende Estriche auf Massivdecken – Richtlinien für die Ausführung. September 1962.
DIN 4109, Blatt 5: Schallschutz im Hochbau; Erläuterungen. März 1963.
DIN 18353 – VOB, Teil C: Estricharbeiten. Oktober 1975.
Bundesverband Estriche und Bodenbeläge (Hrsg.): Technische Richtlinien für die Verlegung schwimmender und im Verbund hergestellter Estriche. Bonn, 1976.


## Substratum and insulating layer

Braun, G.; Ebene Flächen im Hochbau. In: boden – wand – decke, Heft 12/1977, Seite 69–71.
Pohlenz, R.: Schallschutz bei Deckenkonstruktionen. In: deutsche bauzeitung, Heft 3/1977, Seite 40–42.
DIN 1101: Holzwolleleichtbauplatten; Maße, Anforderungen, Prüfung. April 1970.
DIN 18164, Blatt 2: Schaumkunststoffe als Dämmstoffe für das Bauwesen; Dämmstoffe für die Trittschalldämmung. Dezember 1972.
DIN 18165, Blatt 2: Faserdämmstoffe für das Bauwesen; Dämmstoffe für die Trittschalldämmung. Dezember 1972.
DIN 18202, Blatt 3 (V): Maßtoleranzen im Hochbau; Toleranzen für die Ebenheit der Oberflächen von Rohdecken, Estrichen und Oberbelägen. September 1970.
DIN 18202, Blatt 5: Maßtoleranzen im Hochbau; Ebenheitstoleranzen für Flächen von Decken und Wänden. 1979.


## Cement screed

Albrecht, W.; Mannherz, U.: Eigenschaften von Estrichmörteln und schwimmenden Estrichen. In: Boden – Wand – Decke, Heft 9/1966, Seite 742–760; Heft 10/1966, Seite 870–889.
Albrecht, W.; Mannherz, U.: Zusatzmittel, Anstrichstoffe, Hilfsstoffe für Beton und Mörtel. 8. Auflage, Bauverlag, Wiesbaden und Berlin 1968.
Klopfer, H.: Spannungen und Verformungen in Estrichen. In: boden – wand – decke, Heft 4/1978, Seite 81–85; Heft 5/1978, Seite 97–103; Heft 6/1978, Seite 73–74.
Schneider, H.; Schnell, W.; Diem, P.: Untersuchungen über den Austrocknungsverlauf von Estrichen. In: Berichte aus der Bauforschung, Heft 81, Verlag Wilhelm Ernst und Sohn, Berlin 1972.
Zimmermann, G.: Estriche. In: deutsche bauzeitung, Heft 11/1970, Seite 990–996.
Zimmermann, G.: Volumenänderungen von Bauteilen. In: deutsche bauzeitung, Heft 3/1969, Seite 188–196; Heft 5/1969, Seite 350–364; Heft 7/1969, Seite 524–548.
DIN 1045: Beton- und Stahlbetonbau; Bemessung und Ausführung. Januar 1972.


## Damp and wet rooms

Grunau, E. B.: Feuchtigkeitsabdichtungen bei keramischen Bodenbelägen. In: Fliesen und Platten, Heft 2/1975.
Lufsky, K.: Bauwerksabdichtungen; Bitumen und Kunststoffe in der Abdichtungstechnik. 3. Auflage, B. G. Teubner Verlag, Stuttgart 1975.
Rick, A. W.: Abdichtung von Naßräumen. In: boden – wand – decke, Heft 5/1965, Seite 424–428.
Reichert, H.: Sperrschicht und Dichtschicht im Hochbau. Verlagsgesellschaft Rudolf Müller, Köln 1974.

DIN 4122: Abdichtung von Bauwerken gegen nichtdrückendes Oberflächenwasser und Sickerwasser mit bituminösen Stoffen, Metallbändern und Kunststoff-Folien-Richtlinien. Juli 1968.
DIN 18337 – VOB, Teil C: Abdichtung gegen nichtdrückendes Wasser. Februar 1961.


## Floor slab in contact with ground

Probst, R.: Bodenbeläge und Estriche. In: Das Bauzentrum, Heft 3/1969, Seite 77–81.
Rick, A. W.: Decken über nicht unterkellerten Räumen. In: boden – wand – decke, Heft 3/1965, Seite 218–222.
Schild, E.; Oswald, R.; Rogier, D.; Schweikert, H.: Bauschäden im Wohnungsbau, Teil VI; Bauschäden an Kellern, Dränagen und Gründungen – Ergebnisse einer Umfrage unter Bausachverständigen. Verlag für Wirschaft und Verwaltung H. Wingen, Essen 1977.
Schild, E.; Oswald, R.; Rogier, D.; Schweikert, H.: Schwachstellen, Band III – Keller, Dränagen. Bauverlag, Wiesbaden und Berlin 1978.
Schütze, W.: Estriche und Abdichtungen gegen nichtdrückendes Wasser. In: boden – wand – decke, Heft 2/1965, Seite 108–110.


## Underfloor heating

Gerth, D.: Heizungsrohre in der Deckenkonstruktion. In: Heizung, Lüftung, Haustechnik, Heft 6/1967, Seite 209–216.
Metzsch, von, F. A.: Keramische Fliesen auf Bodenheizungen. in: boden – wand – decke, Heft 5/1976, Seite 36–42.
Schütze, W.: Ausführung und Vergütung besonderer Maßnahmen bei Rohren unter schwimmenden Estrichen. In: boden – wand – decke, Heft 12/1966, Seite 1100–1115.


## Thermal insulation

Zimmermann, G.: Aufgestelzter Fußboden auf Stahlbetondecke über offener Durchfahrt. Tauwasserbildung auf der Stahlbetondecke, DAB Bauschäden Sammlung. In: Deutsches Architektenblatt, Heft 6/1976, Seite 486.
Ergänzende Bestimmungen zur DIN 4108: Wärmeschutz im Hochbau (1969). Oktober 1974.
Verordnung über einen energiesparenden Wärmeschutz bei Gebäuden vom 11. August 1977.
Bautechnische Informationen – Bauen mit Naturwerksteinen. Informationsstelle Naturwerkstein, Würzburg, o. J.


## Bonded screeds

– Specialist books and guidelines

Albrecht, W.; Mannherz, U.: Zementmittel, Anstrichstoffe, Hilfsstoffe für Beton und Mörtel. 8. Auflage, Bauverlag, Wiesbaden und Berlin 1968.
Bundesverband Estriche und Bodenbeläge: Technische Richtlinien für die Verlegung schwimmender und im Verbund hergestellter Estriche. Bonn, 1976.
Forschungsgemeinschaft Bauen und Wohnen: Estriche im Hochbau. Heft 80. Verlagsgesellschaft Rudolf Müller, Köln 1967.
Forschungsgemeinschaft Bauen und Wohnen: Estriche im Industriebau. Verlagsgesellschaft Rudolf Müller, Köln 1976.
Seidler, P. (Hrsg.): Industriefußböden. Kontakt + Studium, Band 20. Technische Akademie Esslingen, Lexika-Verlag, Grafenau 1978.
Schild, E.; Oswald, R.; Rogier, D.; Scheikert, H.: Bauschäden im Wohnungsbau, Teil VI; Bauschäden an Kellern, Dränagen und Gründungen – Ergebnisse einer Umfrage unter Bausachverständigen. Verlag für Wirtschaft und Verwaltung H. Wingen, Essen 1977.
Schild, E.; Oswald, R.; Rogier, D.; Scheikert, H.: Schwachstellen, Band III; Keller, Dränagen. Bauverlag, Wiesbaden und Berlin 1978.
Schütze, W.: Kunstharz-Estrich; Beschichten, Fugen, Sanieren. Bauverlag, Wiesbaden und Berlin 2978.
DIN 1045: Beton- und Stahlbetonbau; Bemessung und Ausführung. Januar 1972.
DIN 18202, Blatt 3 (V): Maßtoleranzen im Hochbau; Toleranzen für die Ebenheit der Oberflächen von Rohdecken, Estrichen und Oberbelägen. September 1970.

Floors
Typical cross section

DIN 18202, Blatt 5: Maßtoleranzen im Hochbau; Ebenheitstoleranzen für Flächen von Decken und Wänden. 1979.
DIN 18353 – VOB, Teil C: Estricharbeiten. Oktober 1975.

- Specialist papers

Klopfer, H.: Spannungen und Verformungen in Estrichen. In: boden – wand – decke, Heft 4/1978, Seite 81–85; Heft 5/1978, Seite 97–103; Heft 6/1978, Seite 73–74.
Kranz, S.: Kunstharzmodifizierte Industrieestriche, Imprägnierungen, Versiegelungen, Beschichtungen. In: bau + bauindustrie, Heft 8/1968, Seite 558–565.
Pilny, Fr.: Entstehen und Beherrschen von Beanspruchungen in Plattenbelägen. In: boden – wand – decke, Heft 6/1967, Seite 452–469; Heft 7/1967, Seite 521–530.
Schütze, W.: Kunstharzhaftbrücken für Estriche. In: boden – wand – decke, Heft 12/1975, Seite 77–80.

## Surface coverings

### Tiling

Cantner, M.: Kombiniertes Isolier- und Klebeverfahren für keramische Fliesen. In: Deutsche Bauzeitschrift (DBZ), Heft 1/1975, Seite 79–80.
Grunau, B. B.: Feuchtigkeitsabdichtung bei keramischen Bodenbelägen. In: Fliesen und Platten, Heft 2/1975.
Niemer, E. U.: Dehnungsfugen in keramischen Bekleidungen und Belägen. In: Fliesen und Platten, Heft 4/1977.
Oswald, R.: Schäden an Oberflächenschichten von Innenbauteilen. In: Forum Fortbildung Bau 9, Forum Verlag, Stuttgart 1978.
Pilny, Fr.: Entstehen und Beherrschen der Beanspruchungen in Plattenbelägen. In: boden – wand – decke, Heft 6/1967, Seite 452–458 und Heft 7/1967, Seite 521–530.
Pilny, Fr.; Schröder, H.: Die Beanspruchung keramischer Bodenplatten durch die Verbundwirkung mit dem Untergrund. In: Der Bauingenieur, Heft 2/1969, Seite 37–46.
Probst, R.: Bodenbeläge und Estriche. In: Das Bauzentrum, Heft 3/1969, Seite 77–81.
Pröpster, H.: Schadensanalysen bei Fliesen- und Plattenbelägen. Verlagsgesellschaft Rudolf Müller, Köln-Braunsfeld 1978.
ohne Verfasserangabe: Druckfest Fliesenböden. In: Fliesen und Platten, Heft 21/1973.
DIN 18155 (V), Teil 1–4, Feinkeramische Fliesen, März 1976.
DIN 18156, Teil 1: Stoffe für keramische Bekleidungen im Dünnbettverfahren, Begriffe und Grundlagen. April 1977.
DIN 18156, Teil 2: Stoffe für keramische Bekleidungen im Dünnbettverfahren, Hydraulisch erhärtende Dünnbettmörtel. März 1978.
DIN 18166, Keramische Spaltplatten. Oktober 1974.
DIN 18352 – VOB, Teil C: Fliesen- und Plattenarbeiten. August 1974.
DIN 18332 – VOB, Teil C: Naturwerksteinarbeiten. August 1974.
DIN 18333 – VOB, Teil C: Betonwerksteinarbeiten. August 1974.
Richtlinien für das Versetzen und Verlegen von Naturwerksteinen. Deutscher Naturwerkstein-Verband e.V., Würzburg, 1979.
Empfehlungen für die Anwendung glasierter keramischer Bodenbelagsmaterialien. Fliesen-Beratungsstelle, Großburgwedel, 1976.
Merkblätter der Fliesen-Beratungsstelle. Großburgwedel, 1969.
Bautechnische Informationen – Bauen mit Naturwerkstein. Informationsstelle Naturwerkstein, Würzburg, o. J.

### Parquet flooring

Arbeitsgemeinschaft Holz: Fertigparkettelemente. Informationsdienst Holz.
Koob, H. K.: Parkett- und Holzpflasterfußböden. In: Deutsche Bauzeitschrift, Heft 8/1971, Seite 1561–1570.
Ruske, W.: Fertigparkett. In: Deutsche Bauzeitschrift, Heft 2/1977, Seite 217.
Schellenberg, K.: Mosaikparkett auf Zementestrich; Aufwölben des Parketts, Risse im Estrich. In: Deutsches Architektenblatt, Heft 2/1974, Seite 241–242.
Zimmermann, G.: Mosaikparkett auf Zementestrich; Ablösungen des Mosaikparketts infolge mangelhafter Verklebung und Einwirkung von Wasser. In: Deutsches Architektenblatt, Heft 7/1971, Seite 263–264.
Zimmermann, G.: Volumenänderungen von Bauteilen. In: deutsche bauzeitung, Heft 3/1969, Seite 188–196; Heft 5/1969, Seite 350–364; Heft 7/1969, Seite 524–548.
Zimmermann, G.: Elastischer Sportboden in Mehrzweckhalle; Quellung und Feuchtigkeitsminderung der Holzspanplatten. In: Deutsches Architektenblatt, Heft 4/1977.
DIN 280, Blatt 1: Parkett; Parkettstäbe und Tafeln für Tafelparkett. Dezember 1970.
DIN 280, Blatt 3: Parkett; Parkettriemen. Dezember 1970.
DIN 280, Blatt 4: Parkett; Parkettdicken, Parkettplatten. Juni 1973.
DIN 280, Blatt 5: Parkett; Fertigparkett-Elemente. Juni 1973.
DIN 281: Parkettklebstoff kalt streichbar; Anforderungen, Prüfung. Dezember 1973.
DIN 18356 – VOB, Teil C: Parkettarbeiten. August 1974.

## Problem: Joints and connections

Careful design and construction of detailing, such as joints, perimeter connections and openings are of great importance, especially in the case of floating screeds, screeds laid on insulating layers, and floorcoverings, since the freedom from defects and serviceability of the whole floor structure are largely dependent on them.

Special attention must be paid to these points, since they are responsible for sound and thermal insulation and partly responsible for ensuring that the floor is water-proof; at the same time, they must also allow for expansion as a result of deformation and expansion of the screed and/or the floorcovering.

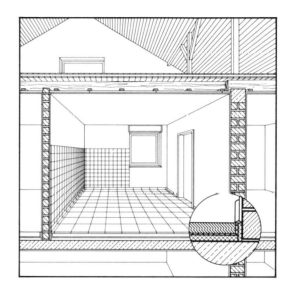

However, since too little attention is paid to these problems in many cases, these points of detail represent marked defects in floors. The effects of defective construction can, for example, reduce the designed high level of impact sound insulation in a floating screed because of acoustic bridging. Similarly, the absence of a flawless seal around the perimeter of a floor slab which is in contact with the ground can cause damp penetration in the floorcovering and the floating screed. Failure to use a suitable expansion joint around the perimeter of the floor surface can produce cracks and warping in the screed and surface covering, especially in the case of an irregular surface structure or underfloor heating.

This and other recurring faults and defects are described and analysed below. Finally, recommendations are made for preventing them.

Floors
Points of detail

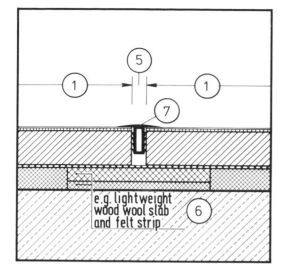

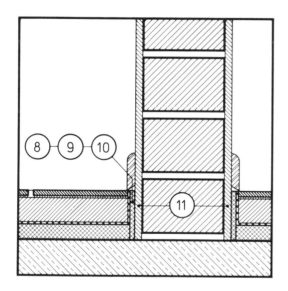

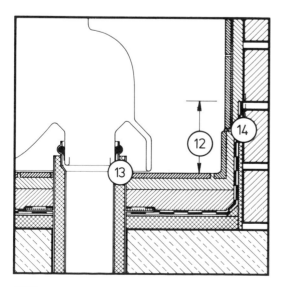

**1** Floating screeds, or screeds laid on an insulating layer must have shrinkage joints at least every 6 m; however, if they are exposed to severe temperature fluctuations, they must be subdivided by expansion joints at maximum intervals of 5 m (see C 2.1.2 and C 2.1.3).

**2** Movement joints in the floor supporting structure must continue through all layers of the floor (see C 2.1.2).

**3** Shrinkage joints can be constructed in the form of narrow 'fake joints' (see C 2.1.2).

**4** Bonded screeds laid on base slabs made of individual sections, as well as screeds on thin insulating layers, should be subdivided by means of 'fake joints' above joints between the individual sections (see C 2.1.2).

**5** The screed and floorcovering must be subdivided by expansion joints which may not be filled with hard materials. The width of these joints depends on the expansion which occurs; it should not, however, be less than 10 mm (see C 2.1.2).

**6** A layer of insulating material at least 200 mm wide should be laid beneath movement and expansion joints in floating screeds to distribute loads. In movement joints, it is also possible to fit round steel dowels into the screed across the joint; one half of the dowel is then inserted into a sleeve or is lubricated (see C 2.1.2).

**7** Movement and expansion joints must be covered. This can be done, for example, with rubber mouldings (mushroom shaped or tubular) or with suitable displaceable mouldings made of metal or plastic (see C 2.1.2).

**8** The floating screed and all tile or parquet floorcoverings must be separated from all structures which penetrate them, such as columns and pipes, by means of a continuous joint or with a vertical strip of insulating material which is completely covered by the skirting (see C 2.1.3).

**9** Flexible materials over 5 mm thick, such as polystyrene foam, are suitable as insulating layers around the perimeter of the floor. The insulating strips and cover should first be fitted so that their tops are above floor level; they should then be trimmed (see C 2.1.3).

**10** In the case of screeds above underfloor heating systems, the perimeter joint must take the form of an expansion joint to accommodate the degree of expansion anticipated. It must be at least 10 mm wide (see C 2.1.3).

**11** Walls made of porous blocks with hollow cavities must be plastered to where the wall is supported on the base slab (see C 2.1.3).

**12** In damp and wet rooms, the damp-proof layer must be continued above the floorcovering and attached to all rising structures, such as walls. In damp rooms, this vertical section of the sealing course must be approx. 50 mm high, whilst in wet rooms it should measure 150 mm (see C 2.1.4).

**13** Any structures which penetrate the damp-proof layer, such as pipes, should wherever possible be laid in conduits which project beyond the surface of the floorcovering by at least 50 mm. A water-proof connection should then be made between the conduit and the sealing layer (see C 2.1.4).

Floors
Points of detail

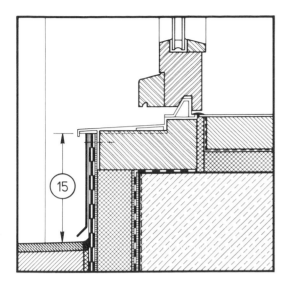

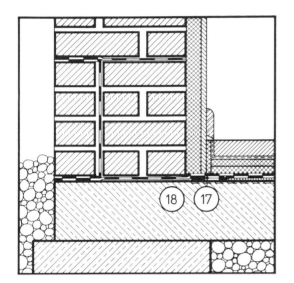

**14** The floorcovering and the wall surface must be separated by an expansion joint. Where possible, this joint should be above any level which is directly exposed to water and must be filled with a mastic sealing compound (see C 2.1.4).

**15** In the threshold areas of room height doors and windows of outdoor areas, the sealing layer must be continued for approximately 150 mm above the level of the floorcovering of the outdoor area. If it is possible to raise the level of the base slab accordingly, the threshold should take the form of a step to the outdoor area. Otherwise, a threshold of suitable height must be built (see C 2.1.5).

**16** The use of thresholds or steps can only be dispensed with if the door is protected against direct exposure to water, e.g. by setting it back inside the level of the wall and at the same time by ensuring that the floor surface falls away from the door (see C 2.1.5).

**17** Along all external and internal walls, the surface damp-proof layers of floor slabs which are in contact with the ground must be continued to the horizontal wall sealing course (see C 2.1.6).

**18** If there is exposure to standing water (basement floors) the seal and the connection must be water-proof; in the event of exposure to spray water (ground floor), the seal must be capillary tight (see C 2.1.6).

Floors
Joints and connections

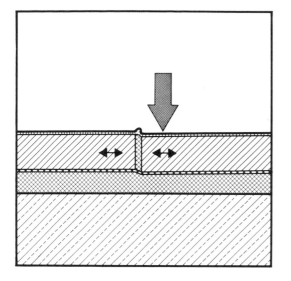

A very large proportion of the cracks in floating screeds and tiled floorcoverings, and creases in thin floorcoverings, could be attributed to the absence of joints or incorrectly constructed joints. The cracks generally appeared directly after the screed had been constructed, although in some cases, it took some time for them to appear – for example, after an underfloor heating system had been taken into service. These defects started at projecting wall corners and door frames, or subdivided large areas of floor.

In screeds laid directly on the unfinished slab, cracks appeared above joints between the supporting slab structure. Cracks like those described in the section on floating screeds were also found if the screed was laid on separating layers or if they were not adequately bonded to the substratum.

## Points for consideration

– The way in which joints in floors are constructed depends primarily on the anticipated levels of deformation and expansion. Thus, a distinction must be made between shrinkage joints, expansion joints, and joints above movement joints in the supporting floor structure.

– In the case of screeds, the particular risk areas are along the perimeter as well as near joints which continue through the whole floor structure. The adhesion surface of bonded screeds is exposed to particularly high shearing stresses in a narrow area around the perimeter which can lead to progressive parting of the screed from the rest of the floor structure. In screeds which are not bonded to the substratum, the perimeter tends to be weakened because the surface dries more quickly than the rest, and these edges may break away completely, merely under their own weight, or when walked upon. Screeds that are laid on flexible layers of insulating material tend to bend much more severely along the walls than in the centre. In addition to the risk of breaking, the lifting of screed edges alongside joints can be damaging. The principle is, thus, to provide all the necessary joints in the screed, but no superfluous ones.

– Shrinkage joints are designed to prevent haphazard cracking in the screed, for example as a result of the initial contraction of the screed. The joint can therefore be very thin, taking the form of a trowel cut. Moreover, if the trowel cut is made only about 1/2 to 2/3 of the depth of the screed, it nevertheless continues to be effective as a shrinkage joint, whilst the remaining connection between the rest of the cross section allows for a certain amount of keying between the separated screed surfaces, thus largely preventing breakage and lifting when a load is applied to one side. Once all shrinkage is complete these joints can, if necessary, be permanently filled.

– If a floating floor, or a floor laid on insulating layers is exposed to severe temperature fluctuations, both the screed and the floorcovering undergo fluctuations in length. Compressive stresses and possible damage can be avoided if joints of sufficient width are made at suitable intervals across and round the perimeter of the screed and the floorcovering, as well as if any other hard layers (levelling screed) above the insulating layer are separated.

– In movement joints in load-bearing structures, variations in length and position of varying degrees can occur in all three directions. At these points, therefore, the movement joint must subdivide all the layers of the floor structure.

Floors
Joints and connections

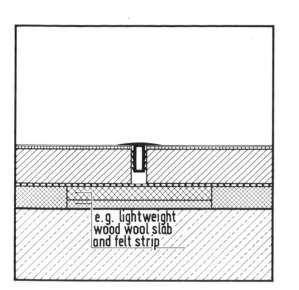

e.g. lightweight
wood wool slab
and felt strip

– Screeds bonded to the base slab or to thin insulating layers are directly affected by deformation or cracks in the supporting structure. Further cracking can be avoided by arranging trowel cuts above joints between individual elements of the supporting structure.

– Expansion and movement joints continuously have to absorb alternating movements and should therefore not be filled with hard materials which would prevent this action. The width of the joints must be compatible with the movements which occur, and with the design and construction of the supporting structure. The minimum width should be at least 10 mm, with the edges of the joints level and parallel – if necessary, they should be covered with metal protecting strips. In some cases, it may be necessary to have a type of cover over the joints which will permit objects to be rolled over them; at the same time, this will prevent damage to the edges of the joint as well as keeping dirt or cleaning water out. In the majority of cases, it is not adequate merely to fill the joints with mastic sealing compounds.

### Recommendations for the avoidance of defects

● Floating cement screeds, or screeds laid on insulating layers must have shrinkage joints at least every 6 m.

● Shrinkage joints can be constructed in the form of narrow 'fake joints' by inserting into the fresh screed a metal strip of conical cross-section which is later removed, or by making a trowel cut in the screed down to no more than 2/3 of its thickness.

● Movement joints in the floor supporting structure must continue through all layers of the floor. They must not be filled with hard materials.

● Bonded screeds laid on base slabs made of individual sections, as well as screeds laid on thin insulating layers, should be subdivided by means of 'fake joints' above joints between the individual sections.

● Floating screeds, or screeds laid on a separating layer, which are exposed to severe temperature fluctuations must be subdivided by expansion joints at maximum distances of 5 m. Similarly, all joints around the perimeter of the screed must take the form of expansion joints.

● Expansion joints must separate the screed and the floorcovering and must not be filled with hard materials. Their width depends on the variations in length which occur; it should not, however, be less than 10 mm.

● In floating screeds laid on layers of insulating material, it is necessary to take steps to avoid lifting in the area of the joint. In the case of movement or expansion joints, this can be achieved, for example, by laying a strip of insulating material at least 200 mm wide beneath the joints to distribute the loads (provided that the use of this strip does not result in an unacceptable reduction in the level of impact sound insulation). In movement joints, it is also possible to fit round steel dowels into the screed across the joint; one half of the dowel is then inserted into a sleeve or is lubricated.

● Movement and expansion joints also separate the floorcovering. They must therefore be covered in such a way that the surface is level and of satisfactory appearance, the edges of the joint are protected from damage, and no moisture or dust can enter the joint. This can be done, for example, with rubber mouldings (mushroom shaped or tubular) or with suitable displaceable mouldings made of metal or plastic.

Floors
Joints and connections

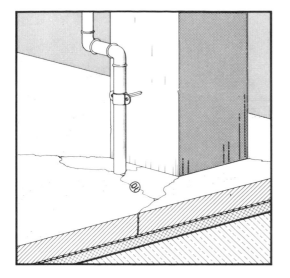

The junction of floating screeds to adjacent structures or to structures which pass through them (e.g. walls, columns or chimneys, pipes, door frames or door thresholds, doormat recesses, etc.) was defective in many cases, thus resulting in damage. Cracks emanating from projecting wall corners, door frames and pipes appeared in the screed if the screed was in direct contact with adjacent structures, or if narrow joints alongside these structures had been filled with rigid materials, e.g. bituminous felts, PVC coverings, etc.

In the above mentioned cases, as well as where – despite sufficiently wide perimeter joints – the screed came into contact with the substratum at certain points, or where hard tile finishes bedded in mortar were continued to make tight contact with adjacent structures, the impact sound insulation of the floor structure was also defective.

**Points for consideration**

– The high degree of sound insulation provided by a floating screed depends on the fact that the load-distributing screed surface is separated on all sides from adjacent structures, or structures which pass though the surface of the screed, so that direct transmission of impact sound in walls or floor slabs is avoided.

– Connections of the screed or hard surface finishes, to walls, for example, act as 'sound bridges', the effect of which increases proportionally to their number or length. In addition, they considerably reduce the level of high frequency sound insulation.

– The necessary separation is best achieved if, along all points of connection, the layers of insulating and covering material are taken up vertically at the edges to provide a continuous upstand of the same depth as the floor construction. The demands placed on the insulating material for the edge connection in terms of rigidity are less than the demands placed on the insulating layer beneath the screed. Thus, even thin insulating layers are adequate, provided that they are sufficiently flexible.

– Non-plastered internal brickwork walls, and particularly walls made of porous blocks, often have open joints or pores which permit direct transmission of air-borne sound through the wall. If the plaster is continued only as far as a plaster stop above the floor and if a skirting board is fitted in front of this, it can lead to additional air-borne sound transmission between adjacent rooms.

– Screeds made of cement mortar shrink after they are laid. However, further expansion and contraction can occur if the floor is exposed to direct sunlight or in screeds laid over underfloor heating systems.

– Such variations in the length of the screed are impeded by projecting wall corners, and structures which are directly anchored in the floor structure and which penetrate the screed. This results in stresses in the screed which will lead to cracks, particularly if the low tensile strength of the screed is exceeded. In addition to the expansion joints which it may be necessary to construct in the surface of the screed (see C 2.1.2: Joints in screeds), all the other connecting joints will act as expansion joints if they are continuous (from a sound insulation viewpoint – see above), sufficiently wide, and filled with a flexible material.

Floors
Joints and connections

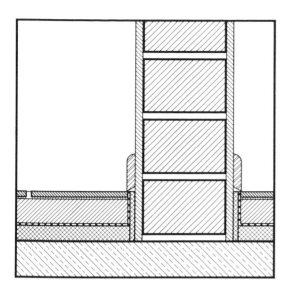

**Recommendations for the avoidance of defects**

● The floating screed must be separated from all adjacent walls and structures which penetrate it, such as columns, pipes, etc. by means of a continuous joint which is separated by a vertical edging strip of insulating material which is completely covered by the skirting.

● Flexible materials over 5 mm thick, such as polystyrene foam, are suitable as insulating layers around the perimeter of the floor. The insulating strips and covering material should first be fitted so that their tops are above floor level; they should then be trimmed.

● If the continuous lengths of the screed exceed 6 m, or if the floor surface is irregular, additional shrinkage joints should be arranged in the floor area (see C 1.1.4: Material selection and size).

● In the case of screeds above underfloor heating systems, the perimeter joint must take the form of an expansion joint to accommodate the degree of expansion anticipated. It must be at least 10 mm wide.

● If the continuous length of the screed exceeds 5 m above underfloor heating systems, or if the floor surface is irregular, additional expansion joints should be arranged in the floor area (see C 1.1.8: Underfloor heating).

● Walls made of porous blocks with hollow cavities must be plastered to where the wall is supported on the base slab.

● Like the screed, mortar-laid tiles or parquet flooring should only continue as far as the perimeter insulating layer in order to prevent a tight connection with the wall.

Floors
Joints and connections

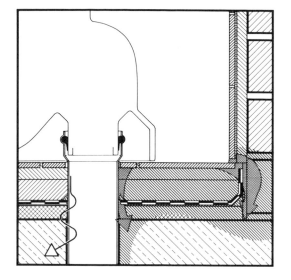

In most cases where damp penetrated insulating layers and adjacent structures beneath bathrooms and shower rooms this could be attributed to incorrectly made connections between the damp-proof layer and adjacent walls or other structures penetrating the floor structure. Thus, for example, the damp-proof layer for the walls continued only to the height of the screed or the tiling.

If the keyed pointing between the floorfinish and the wall surface was tightly packed with mortar, cracking frequently occurred at this point and the impact sound insulation of the floating floor was unsatisfactory.

## Points for consideration

– Repeated exposure to water is to be expected in the case of floors in bath and shower rooms in residential buildings, and particularly in the changing rooms of swimming baths, etc. They may also be exposed to spray if the floor is cleaned with a hose.

– The sealing measures necessary in damp rooms in residential buildings and in wet rooms (see C 1.1.7: Damp and wet rooms) must protect the floor construction and adjacent structures against exposure to water.

– It is also likely that the base of the wall will be exposed to standing water at points where the floorfinish meets the wall, especially where there is no fall to the floor, or where the floor falls towards the walls.

– Depending on the use of the room and the way in which it is cleaned, the base areas of walls can suffer extreme spray water exposure.

– If the joint between the floorfinish and wall tiles is filled with a rigid material it will shear because of the alternating thermal and moisture expansion of the screed and the surface covering. Moreover, if the damp-proof layer to be connected to the walls is continued only as far as the surface of the floor, large quantities of water can penetrate the wall from the floor surface and run behind the sealing layer.

– If tiling is continued directly to an adjacent structure or to one that penetrates the floor, or if the perimeter joint is filled with rigid materials, acoustic bridges are produced which transmit impact sound direct to adjacent walls or ceilings. Moreover, as the number of these bridges increases, the impact sound insulation provided by the floating screed fails, especially at high frequencies.

– The use of bituminous compounds or sealants alone, to connect damp-proof layers at continuous elements such as pipes, will not produce joints which are reliably water-proof and permanent. Suitable connecting structures can only be achieved by means of sufficiently wide bonded flanges or by means of fixed/moving flange connections. Because of the separation of the screed from the tiling, which is necessary because of sound insulation, it is advantageous to install sleeved pipes containing packings which provide insulation against transmitted sound.

Floors
Joints and connections

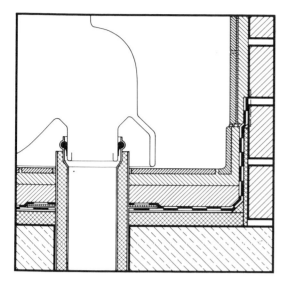

### Recommendations for the avoidance of defects

● The damp-proof layer must be continued above the floorcovering and attached to all rising structures such as walls. In damp rooms in residential buildings, this upstand of the sealing course must be approx. 50 mm high, whilst in wet rooms it should measure 150 mm.

● Any structures which penetrate the damp-proof layer, such as pipes, should wherever possible be laid in conduits. A bonded, fixed or loose flange should then be used to make a water-proof seal between the sealing layer and the conduit, which should project above the floor by at least 50 mm and should be filled with a packing to reduce transmitted sound.

● The floating screed and the floorcovering should be acoustically separated from adjacent structures by means of a continuous joint formed by a vertical strip of insulating material and a cover.

● The floorcovering and the wall surface must be separated by an expansion joint. Where possible, this joint should be above any level which is directly exposed to water and must be filled with a mastic sealing compound.

Floors
Joints and connections

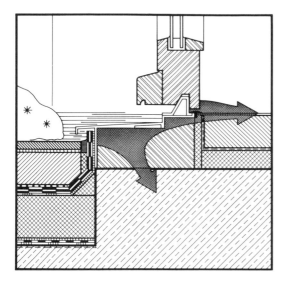

Floors were frequently found to be defective alongside room-height windows (especially in the vicinity of balconies or roof terraces) if the difference in height between the inside and the outside was only slight. This deterioration increased if the floorcovering of the balcony or terrace was higher than that of the internal floor in the area of the threshold, and if the threshold provided was not high enough. In some cases, ponding was visible on the floor, but generally the defects took the form of damage to insulating layers and floorfinishes which were not damp-proof. Parquet flooring lifted from the floor or warped, bonded plastic tiles lifted from the floor and carpets suffered damp patches.

## Points for consideration

- Near the thresholds of room-height windows and doors the floor structure is adjacent to horizontal external surfaces of the building. Thus, special constructional measures are necessary in such cases in order to guard against the same high level of exposure to thawing snow, ponding, spray water and heavy rain experienced by external surfaces.

- The depth of an internal floor, for example a floating screed and a floorfinish, only exceeds 100 mm in exceptional circumstances, whilst at the lowest point of a roof terrace, the roof covering structure including the thermal insulation layer is already in excess of 100 mm. Similarly, because it is desirable to have a fall on balconies, these can also be of approximately this thickness near the threshold.

  If the base slab continues from the inside to the outside on one level, this often results in a situation where the covering for the outdoor area is of the same depth, or indeed is deeper than the floor inside the room.

- The risk of water penetrating the inside of the room is reduced if the seal for the outdoor area is returned to form an upstand which is as high as possible. 150 mm above the surface of the covering has been found to be a minimum upstand.

- With a step structure, this upstand can be achieved if the level of the base slab in the threshold area is lowered on the outside by an amount equivalent to the difference in the depths of the surface coverings on each side plus 150 mm.

  In most examples, however, it is simpler and cheaper to construct a threshold of suitable height in the case of a continuous base slab.

- However, if it is not possible to build a step or a threshold, for example because of wheelchair use, the door and window constructions should be arranged in such a way that they are protected against direct rain and a good fall must be provided away from the threshold to ensure satisfactory drainage.

- For further details on this problem see *Structural failure in residential buildings, Volume I – Flat roofs, roof terraces, balconies.*

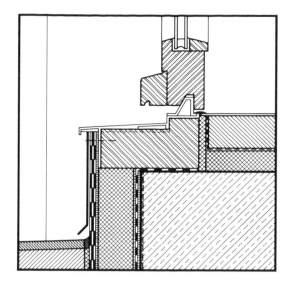

### Recommendations for the avoidance of defects

● In the threshold areas of room-height doors and windows of outdoor areas, the sealing layer must be continued for approximately 150 mm above the level of the floorcovering of the outdoor area.

● If it is possible to raise the level of the base slab accordingly, the threshold should take the form of a step to the outdoor area. Otherwise, a threshold of suitable height must be built.

● The use of thresholds or steps can only be dispensed with if the door is protected against direct exposure to water, e.g. by setting it back inside the level of the wall and at the same time by ensuring that the floor surface falls away from the door.

Floors
Joints and connections

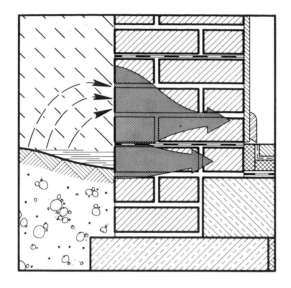

In the case of floors laid on slabs in basements or on the ground floor, damp damage frequently occurred despite damp-proof courses in external and internal walls. Bonded parquet floors lifted from the floor surface and warped, carpets became discoloured, the adhesives for plastic floorcoverings lost their strength thus resulting in blistering and creasing, and insulation layers made of vegetable materials rotted. In all these examples, there was no waterproof connection between the floor and the damp-proof course in the walls' cross-section. In most cases, there were also signs of damp and black fungus on the insides of the walls.

## Points for consideration

– The need for keeping the substratum of floors completely dry varies according to the purpose and construction of the floor and according to the materials used.

– Structures which are adjacent to the ground, such as floor slabs, walls on foundations, etc., are permanently exposed to moisture. Moreover, the base of a wall is also exposed to increased levels of moisture as a result of surface water and spray water.

– If these 'damp' structures are not separated from internal rooms or adjacent structures by means of damp-proof or capillary tight seals, damp penetration damage can occur.

– Where a basement is exposed to groundwater or to water which accumulates, special sealing measures are necessary which protect all structures in the form of 'tanks' (these are described in detail in *Structural failure in residential buildings, Volume III – Basements and adjoining land drainage*).

– Where there is exposure to soil moisture or spray water at the base of a wall, the correct arrangement and connection of seals on the surface of the structures, e.g. the floor slab and the cross-section of the walls, must also ensure a flawless seal between the damp structures and those structures which have to be kept dry.

This can be achieved if a capillary tight joint is made between the seal for the basement floor and the lower horizontal damp-proof course in the external and internal walls, or, in the case of buildings with no basement, with the damp-proof course at the base of the wall.

At the same time, this also prevents distribution of the water through the internal plaster by capillary action.

## Recommendations for the avoidance of defects

● Along all external and internal walls, the surface damp-proof layer of floor slabs which are in contact with the ground must be continued to the horizontal wall-sealing course.

● If there is exposure to standing water (basement floors) the seal and the connection must be water-proof; in the event of exposure to spray water (ground floor), the seal must be capillary tight.

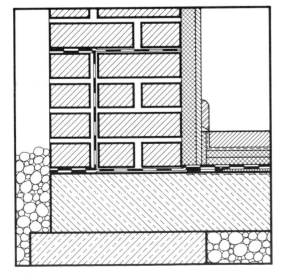

## Specialist books and guidelines

Reichert, H.: Sperrschicht und Dichtschicht im Hochbau. Verlagsgesellschaft Rudolf Müller, Köln-Braunsfeld 1974.

Schild, E.; Oswald, R.; Rogier, D.; Schweikert, H.: Bauschäden im Wohnungsbau, Teil VI; Bauschäden an Kellern, Dränagen und Gründungen; Ergebnisse einer Umfrage unter Bausachverständigen. Verlag für Wirtschaft und Verwaltung H. Wingen, Essen 1977.

Schild, E.; Oswald, R.; Rogier, D.; Schweikert, H.: Schwachstellen, Band III, Keller, Dränagen. Bauverlag, Wiesbaden und Berlin 1978.

Schütze, W.: Der schwimmende Estrich. 4. Auflage, Bauverlag, Wiesbaden und Berlin 1974.

Schütze, W.: Estrichmängel – Entstehen, Vermeiden, Beseitigen. Bauverlag, Wiesbaden und Berlin 1971.

DIN 4109, Blatt 4: Schallschutz im Hochbau; Schwimmende Estriche auf Massivdecken; Richtlinien für die Ausführung. September 1962.

DIN 4109, Blatt 5: Schallschutz im Hochbau; Erläuterungen. April 1963.

DIN 4122: Abdichtung von Bauwerken gegen nichtdrückendes Oberflächenwasser und Sickerwasser mit bituminösen Stoffen, Metallbändern und Kunststoff-Folien; Richtlinien. Juli 1968.

Bautechnische Informationen – Bauen mit Naturstein. Informationsstelle Naturstein. Informationsstelle Naturwerkstein, Würzburg, o.J.

Informationsstelle Naturwerkstein: Bodenbeläge über verschiedenem Bodenaufbau. Bautechnische Informationen 2.7.1972.

## Specialist papers

Hill, G.: Das Verlegen von Fliesen auf Bodenheizungen; Unterkonstruktion, Ausbildung und Anordnung von Dehnungsfugen. In: Haustechnische Rundschau, Heft 9/1976, Seite 402–405.

Oswald, R.: Schäden an Oberflächenschichten von Innenbauteilen. In: Formum Fortbildung Bau 9, Forum Verlag, Stuttgart 1978

Schütze, W.: Estrich-Randfugen. In: boden – wand – decke, Heft 10/1978, Seite 75–76.

Schütze, W.: Das Fugenproblem im schwimmenden Estrich. In: boden – wand – decke, Heft 6/1966, Seite 504–510.

Schütze, W.: Angeschnittene Fugen in Zementestrichen. In: boden – wand – decke, Heft 8/1978, Seite 57–62.

# English Language Bibliography

British Standards Institution, London.
  BS 661: 1969 *Glossary of acoustical terms.*
  BS 882 + 1201: 1973 *Aggregates from natural sources for concrete.*
  BS 1076, 1410 + 1451: 1973 *Mastic asphalt for flooring.*
  BS 1186: Part 1 and 2: 1971 *Quality of timber and workmanship in joinery.*
  BS 1191: Part 1 and 2: 1973 *Gypsum building plasters.*
  BS 1198, 1199 + 1200: 1976 *Building sands from natural sources.*
  BS 1230: 1970 *Gypsum plasterboard.*
  BS 1281: 1974 *Glazed ceramic tiles and tile fittings for internal walls.*
  BS 1286: 1974 *Clay tiles for flooring.*
  BS 1369: 1947 *Metal lathing (steel) for plastering.*
  BS 2592: 1973 *Thermoplastic flooring tiles.*
  BS 2604: 1970 *Resin-bonded wood chipboard.*
  BS 2750: 1956 *Recommendations for field and laboratory measurement of airborne and impact sound transmission in building.*
  BS 4049: 1966 *Glossary of terms applicable to internal plastering, external rendering and floor screeding.*
  BS 4721: 1971 *Ready-mixed lime: sand for mortar.*
  BS 4887: 1973 *Mortar plasticizers.*
  BS 5085: Parts 1 and 2: 1974, 1976 *Backed flexible PVC flooring.*
  BS 5262: 1976 *External rendered finishes.*
  BS 5270: 1976 *Polyvinyl acetate (PVAC) emulsion bonding agents for internal use with gypsum plasters.*
  BS 5386: Part 1: 1976 *Internal ceramic wall tiling and mosaics in normal conditions.*
  BS 5492: 1977 *Internal plastering.*

British Standards Codes of Practice
  CP 3: Chapter III: Part 2: 1972 *Sound insulation and noise reduction (metric units).*
  CP 112: Part 2: 1971 *The structural use of timber (metric units).*
  CP 201: Part 2: 1972 *Timber flooring (board, strip, block and mosaic) (metric units).*
  CP 202: 1972 *Tile flooring and slab flooring.*
  CP 203: Part 2: 1972 *Sheet and tile (cork, linoleum, plastic and rubber).*
  CP 204: Part 2: 1970 *In-situ floor finishes (metric units).*
  CP 209: Part 1: 1963 *Care and maintenance of floor finishes (wood).*

Building Research Establishment, Garston.
  Digest 18 *Design of timber floors to prevent decay.*
  Digest 23 *Damp proof courses.*
  Digest 27 *Rising damp in walls.*
  Digest 33 *Sheet and tile flooring made from thermoplastic binders.*
  Digest 35 *Shrinkage of natural aggregates in concrete.*
  Digest 49 *Choosing specifications for plastering.*
  Digest 54 *Damp proofing solid floors.*
  Digest 55 Part 1 *Painting walls.*
  Digest 75 *Cracking in buildings.*
  Digest 77 *Adhesives used in buildings.*
  Digest 79 *Clay tile flooring.*
  Digests 102 and 103 *Sound insulation of traditional dwellings.*
  Digest 104 *Floor screeds.*
  Digest 142 *Fill and hardcore.*
  Digest 145 *Heat losses through ground floors.*

Department of the Environment
Condensation in dwellings Parts 1 and 2 London HMSO.
  Advisory leaflet 1 *Painting new plaster and cement.*
  Advisory leaflet 2 *Gypsum plaster.*
  Advisory leaflet 5 *Laying floor screeds.*
  Advisory leaflet 6 *Limes for building.*
  Advisory leaflet 9 *Plaster mixes.*
  Advisory leaflet 21 *Plastering on building boards.*
  Advisory leaflet 47 *Dampness in buildings.*
  Advisory leaflet 64 *Plasterboard dry linings.*

Duell, J. and Lawson, F. (1977) *Damp-proof course detailing.* Architectural Press.
Elder, A. J. and Vandenberg, M. (1974) *Architect's Journal Handbook of building enclosure.* Architectural Press.
Eldridge, H. J. (1976) *Common Defects in Buildings.* London HMSO.
King, H. and Everett, A. (1971) *Components and Finishes.* Batsford.
Martin, D. (Ed) (1980) *Specification Vol. 3.* Architectural Press.
Miller, H. (Ed) (1976) *Guide to the choice of wall and floor surfacing materials.* Hutchinson.
Rich, P. (1977) *Principles of Element Design.* Godwin.
Scott, G. (1976) *Building disasters and failures.* A practical report. Construction Press.
Stagg, W. D. and Pegg, B. F. (1976) *Plastering: A craftsmen Encyclopaedia.* Granada Publishing.
Van den Branden, F. and Hartnell, T. O. (1972) *Plastering skills and Practice.* Technical Press.

# Index